LOGIC SYNTHESIS
FOR LOW POWER VLSI DESIGNS

LOGIC SYNTHESIS FOR LOW POWER VLSI DESIGNS

by

Sasan Iman
Escalade Co.
&
University of Southern California

and

Massoud Pedram
University of Southern California

KLUWER ACADEMIC PUBLISHERS
Boston / Dordrecht / London

Distributors for North America:
Kluwer Academic Publishers
101 Philip Drive
Assinippi Park
Norwell, Massachusetts 02061 USA

Distributors for all other countries:
Kluwer Academic Publishers Group
Distribution Centre
Post Office Box 322
3300 AH Dordrecht, THE NETHERLANDS

Library of Congress Cataloging-in-Publication Data

A C.I.P. Catalogue record for this book is available
from the Library of Congress.

Printed on acid-free paper.

Printed in the United States of America

To our loving families.

Contents

Preface

In the past, the major concerns of the VLSI designers were area, speed, cost, and reliability. In recent years, however, this has changed and, increasingly, power is being given comparable weight to area and speed. This is mainly due to the remarkable success of personal computing devices and wireless communication systems, which demand high-speed computation and complex functionality with low power consumption. In addition, there exists a strong pressure for manufacturers of high-end products to keep power under control. The main driving factors for lower power dissipation in these products are the costs associated with packaging and cooling, and circuit reliability.

Although some low power design techniques can be applied manually, the complexity of today's chips is such that tools for hierarchical design capture and automatic design of low-power VLSI circuits are mandatory. The low power design challenge is one that requires abstraction, modeling and optimizations at all levels of design hierarchy, including the technology, circuit, layout, logic, architectural, and algorithmic levels. Combining optimizations at all these levels easily results in orders of magnitude of power reduction. Such impressive reduction in circuit power will however be possible only if optimization flows and techniques at each level of design hierarchy are developed.

Logic synthesis has matured as a field to be universally accepted and used in every major IC design and production house worldwide. A wealth of research results and a few pioneering commercial tools for low power logic synthesis have appeared in the last couple of years. It is our experience that optimization at the logic (gate) level provide 50-70% power reduction without sacrificing the circuit speed.

A systematic and comprehensive treatment of power modeling and optimization at logic level is lacking today. It is the intention of the present book to occupy this niche area. More precisely, this book provides a detailed presentation of methodologies, algorithms and CAD tools for power modeling, estimation and analysis, synthesis and optimization at the logic level. The book contains detailed descriptions of technology-independent logic transformations and optimizations, technology decomposition and mapping, and post-mapping structural optimization techniques for low power. It also emphasizes the trade-off techniques for two-level and multi-level logic circuits that involve power dissipation and circuit speed, in the hope that the readers can better understand the issues and ways of achieving their power dissipation goal while meeting the timing constraints.

Much of the algorithms described in this book have been implemented and released as part of the POSE package. Interested readers may obtain a copy of the POSE program from the following site: http://atrak.usc.edu/~pose . We expect this field to remain active in the foreseeable future. New trends and techniques will emerge, some approaches described in this book will solidify, while others will be improved on; this in view of technological and strategic changes in the world of microelectronics.

The book is written for VLSI design engineers, CAD professionals, and students who have had a basic knowledge of CMOS digital design and logic synthesis. Emphasis is given to top-down structured design flow for ASICs. Examples and benchmark results are presented to qualitatively and quantitatively assess the effectiveness of various point tools and techniques used in the overall design flow. The book can also be used as a textbook for teaching an advanced course on low power logic synthesis. Instructors can select various combinations of chapters and augment them with some of the many references included at the end of each chapter and material selected from a standard textbook on logic synthesis (such as the book by Gary D. Hachtel and Fabio Somenzi on Logic Synthesis and Verification Algorithms published by Kluwer Academic Publishers).

We hope that this book will serve as the reference for logic-level power modeling and optimization and will be complemented in the future by similar books written on physical design and system-level design for low power.

Sasan Iman, Santa Clara, California
Massoud Pedram, Los Angeles, California
September 1997

Acknowledgments

The work described in this book is part of an on-going research project at the University of Southern California to develop power-savvy design methodologies and tools for effective power analysis and optimization at RT and logic levels. The project is named POSE which stands for Power Optimization and Synthesis Environment.

We would like to acknowledge a number of colleagues and Ph.D. students involved in this project. Specifically, we thank Chi-ying Tsui for the insightful discussions on power optimization techniques for PLA designs. He also developed the algorithms for technology mapping and decomposition and gate re-sizing as described in chapter 7. Designing and implementing the POSE system was a challenging part of this work. Special thanks go to Chih-Shun Ding and Chun-Li Pu for contributing to the system development.

The weekly seminars organized by the USC Low Power CAD group was instrumental in providing the necessary forum for identifying research challenges and putting new ideas to test. Special thanks go to all members of the group, present and past, whose hard work, academic excellence and diligent participation, were essential in sustaining this research forum. We would like to thank Jui-Ming Chang, Cheng-Ta Hsieh, Yung-Te Lai, Jaewon Oh, Bahman Salehi Nobandegani, Kuo-Rueih Ricky Pan, Diana Marculescu, Radu Marculescu, Amir Shaygan Salek, Hirendu Vaishnav, and Qing Wu.

The research described in this book was sponsored in part by DARPA under contract number F33615-95-C1627, by SRC under contract number 94-DJ-559, and by NSF NYI and PECASE awards.

Part I

Background, Terminology and, Power Modeling

CHAPTER 1

Introduction

With the increasing complexity of computing devices, tools for automatic synthesis of digital systems have become an integral part of the design cycle at all levels of design abstraction. Physical design and logic synthesis tools are indispensable in designing million transistor chips. This trend has been followed by the introduction of high level design automation tools in the past few years. The main goal in developing such tools has been to satisfy demands for even faster and denser chips, while reducing the design cycle time. The success of design automation tools in increasing performance/density, coupled with recent trends in technology and applications for consumer products, has transformed "power consumption" from a minor concern to a major design challenge.

Perhaps the primary driving factor for designing low power systems has been the remarkable success and growth of the class of portable personal computing devices and wireless communication systems which demand complex functionality, high speed communication, low weight, and long operation times before the battery has to be recharged. In these applications average power consumption is a critical design constraint. The projected power consumption for a portable multimedia terminal, when implemented using the off-the-shelf components not optimized for low-power operation, is about 40W. With advanced Nickel-Metal-Hydride (secondary) battery technologies yielding around 65 watt-hours/kilogram [15], this terminal would require an unacceptable 6 kilograms of batteries for 10 hours of operation between recharges. Even with new battery technologies, such as rechargeable Lithium Ion or Lithium Polymer cells, it is anticipated that the expected battery lifetime will increase to about 90-110 watt-hours/kilogram over the next 5 years [15], which still leads to an unacceptable 3.6 to 4.4 kilograms of battery cells. In the absence of low-power design techniques, then current and future portable devices will suffer from either a very short battery life, or a very heavy battery pack (Figure 1.1).

There also exists a strong pressure for producers of high-end products to reduce their power consumption. Contemporary performance optimized microprocessors dissipate as much as 15-30 W at 100-200 MHz clock rates [4]. In the future, it can be extrapolated that a 10 cm^2 microprocessor, clocked at 500MHz, would consume about 300W. The cost associated with packaging and cooling such devices is huge. Since core power consumption must be dissipated through the packaging, increasingly expensive packaging and cooling strategies are required. Unless power consumption is dramatically reduced, the resulting heat will limit the feasible packing and performance of VLSI circuits and systems. Consequently, there is a clear financial advantage in reducing the power consumed in high performance systems.

In addition to cost, there is the issue of reliability. High power systems often run hot; at the same time, high temperatures tend to exacerbate several silicon failure mechanisms. Every 10 degrees increase in operating temperature roughly doubles failure rate for the components [17]. In this context, peak power (maximum possible power consumption) is a critical design factor because it determines the thermal and electrical limits of designs, impacts the system cost, size and weight, dictates specific battery type, component and system packaging and heat sinks, and aggravates the resistive and inductive voltage drop problems. It is, therefore, essential to have the peak power under control.

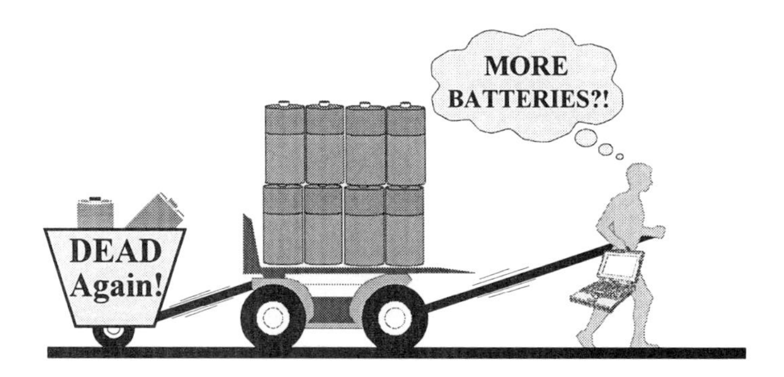

Figure 1.1 Longer battery lifetime is required for portable devices

From the environmental viewpoint, the smaller the power dissipation for electronic systems, the lower the heat pumped into rooms, the lower the electricity consumed and, therefore, the less the impact on the global environment, and the less significant the environment/office power delivery and cooling requirements.

1.1 Low Power Design Methodology

To address the challenge to reduce power, the semiconductor industry has adopted a multifaceted approach, attacking the problem on four fronts:

- **Reducing chip and package capacitance:** This can be achieved through process development such as SOI with fully depleted wells, process scaling to sub-micron device sizes, and advanced interconnect substrates such as Multi-Chip Modules (MCM). This approach can be very effective, but is expensive.
- **Scaling the supply voltage:** This approach can be very effective in reducing the power dissipation, but often requires new IC fabrication processing. Supply voltage scaling also requires support circuitry for low-voltage operation, including level-converters and DC/DC converters, as well as detailed consideration of issues such as signal-to-noise.
- **Using power management strategies:** The power savings that can be achieved by various static and dynamic power management techniques are very application dependent, but can be significant.
- **Employing better design techniques:** This approach promises to be very successful because the investment to reduce power by design is relatively small in comparison to the other approaches, and because it is relatively untapped in potential.

Low power VLSI design can be achieved at various levels of design abstraction from algorithmic and system levels down to layout and circuit levels (Figure 1.2).

At the system level, inactive hardware modules may be automatically turned off to save power. Modules may be provided with the optimum supply voltage and interfaced by means of level converters. Some of the energy that is delivered from the power supply may be cycled back to the power supply. A given task may be partitioned between various hardware modules or programmable processors, or both, so as to reduce the system-level power consumption.

At the architectural (behavioral) level, concurrency increasing transformations, such as loop unrolling, pipelining, and control flow optimization as well as critical path reducing transformations such as height minimization, retiming, and pipelining may be used to allow a reduction in supply voltage without degrading system throughput. Algorithm-specific instruction sets may be utilized that boost code density and minimize switching. A Gray code addressing scheme can be used to reduce the number of bit changes on the address bus. On-chip cache may be added to minimize external memory references. Locality of reference may be exploited to avoid accessing global resources such as memories, busses or ALUs. Control signals that are "don't cares" can be held constant to avoid initiating nonproductive switching.

At the register-transfer (RT) level, symbolic states of a finite state machine (FSM) can be assigned binary codes to minimize the number of bit changes in the combinational logic for the most likely state transitions. Latches in a pipelined design can be repositioned to eliminate hazardous activity in the circuit. Parts of the circuit that do not contribute to the present computation may be shut off completely. Output logic values of a circuit may be precomputed one cycle before they are required and then used to reduce the internal switching activity of the circuit in the succeeding clock cycle.

At the physical design level, power may be reduced by using appropriate net weights during netlist partitioning, floor-planning, placement and routing. Individual transistors may be sized down to reduce the power dissipation along the non-critical paths in a circuit. Large capacitive loads can be buffered using optimally sized inverter chains so as to minimize the power dissipation subject to a given delay con-

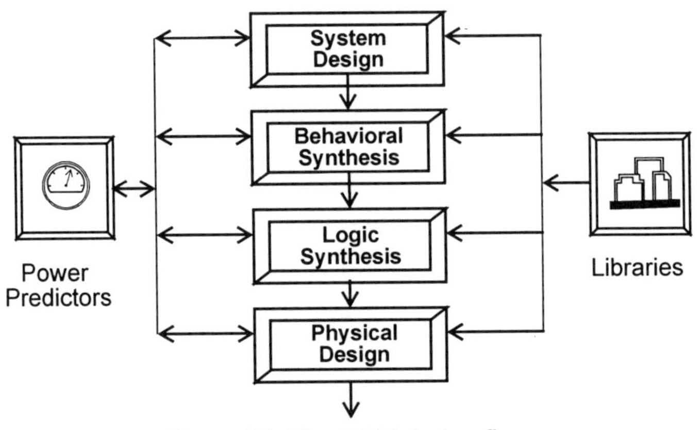

Figure 1.2 The ASIC design flow

straint. Wire and driver sizing may be combined to reduce the interconnect delay with only a small increase in the power dissipation. Clock trees may be constructed that minimize the load on the clock drivers, subject to meeting a tolerable clock skew.

At the circuit level, power saving techniques that recycle the signal energies using the adiabatic switching principles, rather than dissipating them as heat, are promising in certain applications where speed can be traded for lower power. Similarly, techniques based on combining self-timed circuits with a mechanism for selective adjustment of the supply voltage that minimizes the power, while satisfying the performance constraints show good signs of power reduction.

Techniques based on partial transfer of the energy stored on a capacitance to some charge sharing capacitance and then reusing this energy at a later time. Design of energy efficient level-converters and DC/DC converters is also essential to the success of adaptive supply voltage strategies.

Logic synthesis is an important part of the design cycle for a digital system. This means that in order to minimize power effectively, power has to be considered during logic synthesis and optimization. In the following section, an overview of the role of logic synthesis in optimizing a Boolean network will be provided.

1.2 Logic Synthesis for Low Power

Logic synthesis fits between the register transfer level and the netlist of gates specification. It provides the automatic synthesis of netlists minimizing some objective function, subject to various constraints. The goal, in general, is to obtain a minimum area circuit, subject to a given delay requirement. Example inputs to a logic synthesis system include two-level logic representation, multi-level Boolean networks, finite state machines, and technology mapped circuits. Depending on the input specification (combinational versus sequential, synchronous versus asynchronous), the target implementation (two-level versus multi-level, unmapped versus mapped, ASICs versus FPGAs), the objective function (area, delay, power, testability) and the delay models used (zero-delay, unit-delay, unit-fanout delay, or library delay models), different techniques are applied to transform and optimize the original RTL description.

These optimization techniques include the global area minimization strategy (iteratively extract and re-substitute common sub-expressions, eliminate the least useful factors), combinational speed-up techniques based on local restructuring, local optimizations (node factoring and decomposition, two level logic minimization using local don't care conditions), and technology mapping for area or performance optimization.

The basic idea behind extraction techniques is to look for expressions that are observed many times in the nodes of a Boolean network, and extract such expressions. The extracted expression is implemented only once, and the output of that node replaces the expression in any other node in the network where the expression appears. Re-substitution is the process of using the function of a node already in the network to implement the function of another node in the network. Combinational speedup algorithms work by identifying a set of nodes, which when removed reduce the delay of the network. The operation then proceeds to eliminate these nodes from the network by collapsing them into their fanout nodes. Local optimization techniques include two-level node simplification using the local don't care set of the network, which is derived from the network structure, and also the don't care specified by the user. These techniques work on finding a minimal representation of the function of the node. Factorization and decomposition techniques operate by identifying logic sharing within the same function which, when extracted, will result in a smaller representation of the function. Technology mapping is used to implement optimized Boolean network using the gates in the target technology, such as standard cells and FPGAs.

Figure 1.3 shows the typical flow for logic synthesis targeting minimum area subject to a target delay constraint. The approach here is to first optimize the network for area and then use speedup techniques to reduce the network delay. The network is then mapped to the target technology using a technology mapping algorithm. If the delay constraints are not met, the network is then resynthesized to achieve the desired delay requirements.

The power consumption of a Boolean network is in general, directly proportional to the area of the circuit. This means that even though it is possible to make area-power trade-offs, the general trend is that as area is reduced, the power is also reduced. This means that any paradigm for minimizing the power consumption of a Boolean network should also attempt to keep the increase in area to a minimum. Current techniques for minimizing circuit area have proved quite effective. This means that these techniques are a good starting point for developing a power optimization approach. This book will present power optimization techniques which will minimize power by using new approaches, and also by taking advantage of existing techniques for minimizing area. The application of these techniques in the form of an optimiza-

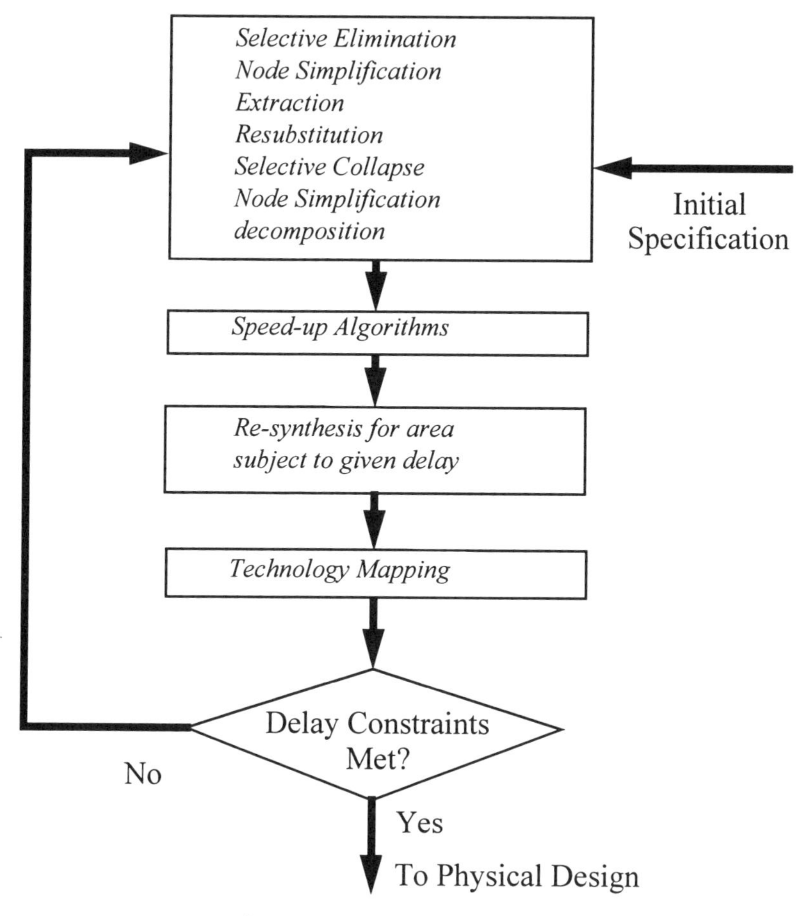

Figure 1.3 Logic Synthesis Flow

tion script for low power introduces new issues that need to be considered. The goal of this book is to provide the methodology and procedures necessary to perform technology independent logic synthesis on a Boolean network, such that, after mapping the power consumption of the network as defined in the next section, is minimized.

The next section provides models for estimating the power consumption of a Boolean network. Based on these models, techniques will be presented which will be used to minimize the network power consumption.

1.3 Sources of Power Dissipation

Power dissipation in CMOS circuits is caused by four sources as follows:

- The *leakage current*, which is primarily determined by the fabrication technology, consists of two components: 1) reverse bias current in the parasitic diodes formed between source and drain diffusions, and the bulk region in a MOS transistor, and 2) the subthreshold current that arises from the inversion charge that exists at the gate voltages below the threshold voltages;
- the *standby current* which is the DC current drawn continuously from v_{dd} to ground;
- the *short-circuit current* (rush-through) current which is due to the DC path between the supply rails during output transitions;
- the *capacitance current* which flows to charge and discharge capacitive loads during logic changes.

The diode leakage is proportional to the area of the source or drain diffusion, and the leakage current density, and is typically 1 *picoA* for a 1 micron minimum feature size. The subthreshold leakage current for long channel devices increases linearly with the ratio of the channel width over channel length, and decreases exponentially with V_{GS}- V_t where V_{GS} is the gate bias and V_t is the transistor threshold voltage. Several hundred milivolts of "off bias" (300-400 *Mv*) typically reduces the subthreshold current to negligible values. With reduced power supply and device threshold voltages, the subthreshold current will however become more pronounced. In addition, at short channel lengths, the subthreshold current also becomes exponentially dependent on drain voltage instead of being independent of V_{DS} [5].

The standby power consumption happens, for example, when both the nMOS and pMOS transistors are continuously on in a pseudo-nMOS inverter, when the drain of an NMOS transistor is driving the gate of another nMOS transistor in a pass-transistor logic, or when the tri-stated input of a CMOS gate leaks away to a value between V_{dd} and ground. The standby power is equal to the product of V_{dd}, and the DC current drawn from the power supply to ground.

The term static power dissipation refers to the sum of leakage and standby dissipations. Leakage currents in CMOS circuits can be made small with the proper choice of device technology. Standby currents are important in CMOS design styles like pseudo-nMOS and nMOS pass transistor logic, and in memory cores.

The short circuit power consumption for an inverter gate is proportional to the input ramp time, the load, and transistor sizes of the gate. The maximum short circuit current flows when there is no load; this current decreases with the load. Depending on the approximations used to model the currents, and to estimate the input signal dependency, different formulae [8] [21], with varying accuracy, have been derived for the evaluation of the short circuit power. A useful formula was recently derived in [19] that shows the explicit dependence of the short circuit power dissipation on the design and performance parameters, such as transistor sizes, input and output ramp times, and the load. The idea is to adopt an alternative definition of the short circuit power dissipation, through an equivalent (virtual) short circuit capacitance C_{SC}.

If gate sizes are selected so that the input and output rise/fall times are about equal, the short-circuit power consumption will be less than 15% of the dynamic power consumption [21]. If, however, design for high performance is taken to the extreme where large gates are used to drive relatively small loads, and if the input ramp time is long, there will be a stiff penalty in terms of short-circuit power consumption.

The dominant source of power dissipation in CMOS circuits is the charging and discharging of the node capacitances (also referred to as the capacitive power dissipation), and is given by:

$$P_i = \frac{1}{2} \cdot V_{dd}^2 \cdot C_i \cdot f \cdot E_i \tag{1.1}$$

where V_{dd} is the supply voltage, f is the clock frequency, C_i is the capacitance seen by gate n_i and E_i (referred as the *switching activity*) is the expected number of transitions at the output of gate g per clock cycle. The product of E_i and f, which is the number of transitions per second, is referred to as the *transition density* [13]. The product of C_i and E_i is referred to as the *switched capacitance*. At the logic level, it is assumed that V_{dd} and f are fixed, and thus, the total switched capacitance of the circuit is assumed to be the cost measure that is optimized.

An estimate for the total power consumption of a network is obtained by summing equation 1.1 over all the nodes in the technology mapped network. Leakage and sub-threshold currents are small compared to charging and discharging currents, and are ignored in this model. As mentioned, short circuit currents can be modeled as an equivalent capacitance which is added to C_i. During logic synthesis, all architectural and technology related issues for the network being optimized have been decided. This means that the values for V_{dd} and f are fixed during logic synthesis. Therefore, the main issues in computing the power estimate are in computing the physical capacitance, and switching activity value at the output of nodes in the Boolean network. These issues are discussed in the next section.

1.4 Physical Capacitance

Power dissipation is dependent on the physical capacitances seen by individual gates in the circuit. Accurate values for this load can be obtained for a mapped network by using the logic and delay information from the target library. Estimating this capacitance at the technology independent phase of logic synthesis is, however, difficult and imprecise since it requires estimation of the load capacitances from structures which are not yet mapped to gates in a cell library.

Meanwhile, interconnect plays a role in determining the total chip area, delay and power dissipation, and hence, must be accounted for as early as possible during the design process. The interconnect capacitance estimation is, however, a difficult task, even after technology mapping, due to the lack of detailed place and route information. Approximate estimates can be obtained by using information derived from a companion placement solution, or by using stochastic/procedural interconnect models [14].

1.5 Switching Activities

Switching activity at the output of a gate gives the number of times in a clock cycle that the value at the output of the node changes its value. Hazards and glitches contribute to the switching activity of a node. Therefore, the value of the switching activity at a node is dependent on the delay model being used.

1.5.1 Delay model

Based on the delay model used, the power estimation techniques could account for steady-state transitions (which consume power, but are necessary to perform a computational task) and/or hazards and glitches (which dissipate power without doing any useful computation). Sometimes, the first component of power consumption is referred as the *functional activity,* while the latter is referred as the *spurious activity.*

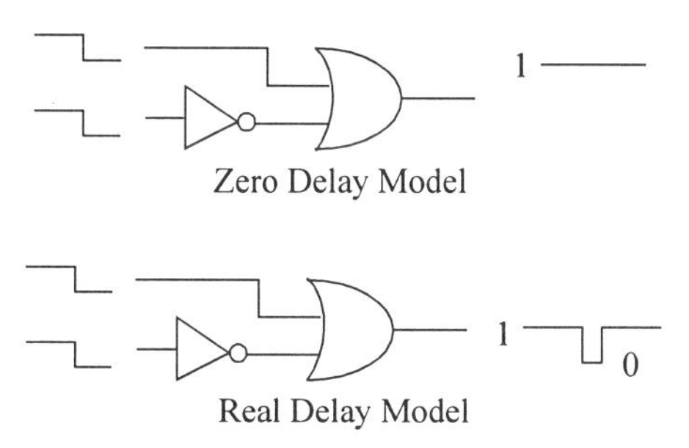

Figure 1.4 Effect of the delay model

It is shown in [1] that the mean value of the ratio of hazardous component to the total power dissipation varies significantly with the considered circuits (from 9% to 38%), and that the spurious power dissipation cannot be neglected in CMOS circuits. Indeed, an average of 15-20% of the total power is dissipated in glitching. The spurious power dissipation is likely to become even more important in the future scaled technologies.

Current power estimation techniques often handle both zero-delay (non-glitch), and real delay models. In the first model, it is assumed that all changes at the circuit inputs propagate through the internal gates of the circuits instantaneously. The latter model assigns each gate in the circuit a finite delay, and can thus account for the hazards in the circuit (Figure 1.4). A real delay model significantly increases the computational requirements of the power estimation techniques, while improving the accuracy of the estimates.

1.5.2 Computing switching activities

Two approaches used for computing switching activities are *simulation based techniques* and *probabilistic techniques*. Circuit simulation based techniques [9],[20] simulate the circuit with a representative set of input vectors. They are accurate and capable of handling various device models, different circuit design styles, single and multi-phase clocking methodologies, tri-state drives, etc. However, they suffer from memory and execution time constraints, and are not suitable for large, cell-based designs. In general, it is difficult to generate a compact stimulus vector set to calculate accurate activity factors at the circuit nodes. The size of such a vector set is dependent on the application and the system environment [16].

The probabilistic techniques use the statistical properties of the circuit inputs, and the structural and functional properties of the network to compute a switching activity for each node. Probabilistic methods for estimating the activity factor $E(n)$ at a circuit node n involve estimation of signal probability $sp(n)$, which is the probability that the signal value at the node is one. Under the assumption that the values applied to each circuit input are temporally independent, (that is, value of any input signal at time t is independent of its value at time t-1), we can write:

$$E(n) = 2 \cdot sp(n) \cdot (1 - sp(n)) \tag{1.2}$$

If a network consists of simple gates and has no re-convergent fanout nodes, then the exact signal probabilities can be computed during a single post-order traversal of the network using the following equations [6]:

NOT gate: $sp(o) = 1 - sp(i)$

AND gate: $sp(o) = \prod_{i \in inputs} sp(i)$

OR gate: $sp(o) = 1 - \prod_{i \in inputs} (1 - sp(i))$

This simple algorithm is known as the *tree algorithm* because it produces the exact signal probabilities for a tree network. For networks with Reconvergant fanout, the tree algorithm, however yields approximate values for the signal probabilities.

In [3], an exact procedure based on Ordered Binary-Decision Diagrams (OBDDs) [2] is described, which is linear in the size of the corresponding function graph (the size of the graph, however, may be exponential in the number of circuit inputs). In this method, which is known as the *OBDD-based* method, the signal probability at the output of a node is calculated by first building an OBDD corresponding to the *global function* of the node (i.e., function of the node in terms of the circuit inputs), and then performing a post-order traversal of the OBDD using equation:

$$sp(y) = sp(x)sp(f_x) + sp(\bar{x})sp(f_{\bar{x}}) \tag{1.3}$$

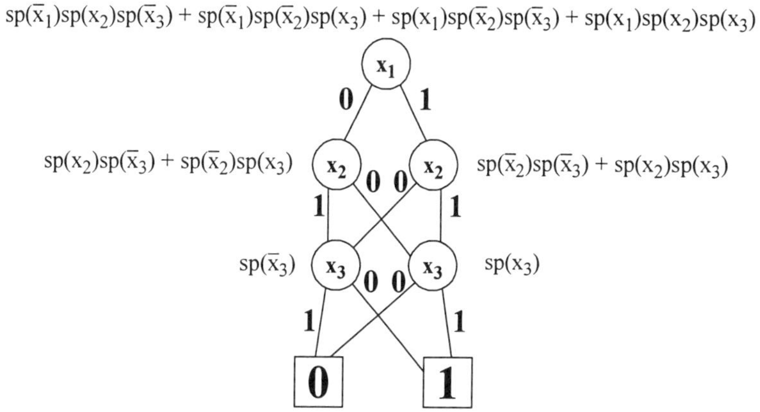

$sp(\bar{x}_1)sp(x_2)sp(\bar{x}_3) + sp(\bar{x}_1)sp(\bar{x}_2)sp(x_3) + sp(x_1)sp(\bar{x}_2)sp(\bar{x}_3) + sp(x_1)sp(x_2)sp(x_3)$

$sp(x_2)sp(\bar{x}_3) + sp(\bar{x}_2)sp(x_3)$

$sp(\bar{x}_2)sp(\bar{x}_3) + sp(x_2)sp(x_3)$

$sp(\bar{x}_3)$

$sp(x_3)$

Figure 1.5 Computing the signal probability using OBDDs

This leads to a very efficient computational procedure for signal probability estimation. Figure 1.5 shows an example computation on the OBDD representation of a three-input **XOR** gate.

Other techniques have also been proposed which provide trade-offs in speed-accuracy by propagating signal probabilities from the circuit inputs toward the circuit outputs, using only *pair-wise correlations* between circuit lines, while ignoring higher order correlation terms [11].

The above methods only account for steady-state behavior of the circuit, and thus ignore hazards and glitches. Techniques have also been developed that examine the dynamic behavior of the circuit, and thus estimate the power dissipation due to hazards and glitches. Such techniques include the symbolic simulation method [7], the probabilistic simulation technique [12], and the tagged probabilistic simulation approach [18].

1.6 Models, Algorithms and Methodologies

Power can effectively be optimized during logic synthesis. It is, however, necessary to guide the optimization process using accurate power estimates during the synthesis process. Significant research has been performed in estimating the switching activity of wires in a digital circuit. Once the switching activity information is available, the power consumption of a technology mapped network can easily be computed based on the description of cells in the library, and using equation 1.1. The problem of power estimation is more complicated in a technology independent Boolean network. The reason is that nodes in a technology independent network do not correspond to any gates in the target library and the final implementation of the circuit after the synthesis process may look different from that of the initial Boolean network. This book describes cost functions which can be used to model the contribution of a node in the unmapped network to the power consumption of the mapped network. It is shown that these cost functions provide accurate estimates of the switched capacitance in the technology mapped network.

As stated before, the switching activity of nodes in a network depends on the functional and spurious activity at the output of that node. Functional activity is dependent on the global function of the node, while the spurious activity is dependent on the network structure. As will be shown in the next chapter, during technology independent optimization, good estimates for the functional activity of nodes in the network can be obtained. However, such estimates for spurious activity cannot be obtained with any degree of accuracy. This book concentrates on minimizing power, assuming a zero-delay model for calculating the switching activity values in the technology independent network. Even though the gain obtained in power consumption of the network by minimizing power assuming a zero-delay model may be reduced due to spurious activities, this will not drastically affect the final results since spurious activity on average accounts for less than %20 of network power. It is also important to note that the steps performed for minimizing logic has a randomizing effect on the spurious activity in the network. In general, for the same circuit, the ratio of the spurious activity over functional activity remains the same as the design is synthesized for different constraints. Therefore, by reducing zero-delay model power consumption, the spurious activity is also reduced to a first order. At the same time, spurious activity can more effectively be minimized during post-mapping optimization steps.

Optimization techniques and algorithms are discussed in this book in the context of two-level function optimization and multi-level network synthesis. Techniques described for two-level functions include optimizations for both dynamic and static CMOS circuits. Multi-level optimization techniques include technology independent optimization, technology mapping, and post-mapping optimization for low power.

1.7 Outline of the Book

This book describes the models, algorithms, and methodologies which are required for effective power minimization during logic synthesis. Chapter 1 discussed the motivation for low power systems, and specifically, the need to consider power optimization during logic synthesis. This chapter also described sources of power consumption and optimization goals that need to be targeted during logic synthesis, namely, reducing switched capacitance. Chapter 2 introduces a technology independent power model that will be used during the technology independent phase of logic synthesis for low power. Chapter 3 describes an exact two-level function optimization algorithm which targets CMOS implementations. This chapter also provides an analysis of the complexity of this operation compared to exact two-level function optimization for minimum area. Chapter 4 provides an exact two-level function optimization for low power that targets PLA implementations of Boolean function. These implementations cover pseudo-NMOS and dynamic PLAs. Chapter 5 provides a complete set of algebraic restructuring techniques to minimize the power consumption of a Boolean network. These operations include common sub-function extraction, factorization, decomposition and selective elimination. Chapter 6 illustrates that don't care computation techniques for minimum area cannot be used freely during power optimization. It then presents a new set of techniques for computing power conscious satisfiability and observability don't cares. Using the computed don't care, it then develops a new function optimization technique based on minimal variable supports that targets multi-level implementation of a Boolean function. Chapter 7 describes technology dependent optimization techniques for low power. Chapter 8 describes a post-mapping power optimization technique based on structural transformations of the netlist of gates. Techniques for performing structural transformation are presented and approaches for speeding up the optimization procedure are discussed. Chapter 9 introduces *POSE, the Power Optimization and Synthesis Environment* and presents a methodology for performing power optimization. The final chapter, 10, concludes the book and proposes avenues for further research.

References

[1] L. Benini, M. Favalli, and B. Ricco. “Analysis of hazard contribution to power dissipation in CMOS ICs.” In proceedings of the *International Workshop on Low Power Design*, pages 27–32, April 1994.

[2] R. Bryant. “Graph-based algorithms for Boolean function manipulation.” In*IEEE Transactions on Computers*, volume C-35, pages 677–691, August 1986.

[3] S. Chakravarty. “On the complexity of using BDDs for the synthesis and analysis of Boolean circuits.” In proceedings of the *27th Annual Allerton Conference on Communication, Control and Computing*, pages 730–739, 1989.

[4] D. Dobberpuhl et el, “A 200MHZ, 64b, dual issue CMOS microprocessor.” Digest of Technical papers, *ISSC '92*, pages 106-107, 1992.

[5] T. A. Fjeldly and M. Shur. " Threshold voltage modeling and the subthreshold regime of operation of short-channel MOSFET's. " *IEEE Transactions on Electron Devices*, 40(1):137–145, Jan. 1993.

[6] H. Goldstein. “Controllability/observability of digital circuits.” In *IEEE Transactions on Circuits and Systems*, 26(9):685–693, September 1979.

[7] A. Ghosh, S. Devadas, K. Keutzer, and J. White. “Estimation of average switching activity in combinational and sequential circuits.” In proceedings of the*ACM/IEEE Design Automation Conference*, pages 253–259, June 1992.

[8] N. Hedenstierna and K. Jeppson. " CMOS circuit speed and buffer optimization. " *IEEE Transactions on Computer-Aided Design of Integrated Circuits and Systems*, 6(3):270-281, March 1987.

[9] S. M. King. “Accurate simulation of power dissipation in VLSI circuits.” In*IEEE Journal of Solid State Circuits*, 21(5):889–891, Oct. 1986.

[10] R. Lisanke, editor. “FSM benchmark suite,” Microelectronics Center of North Carolina, Research Triangle Park. North Carolina, 1987.

[11] R. Marculescu, D. Marculescu, M. Pedram. “Logic level power estimation considering spatio-temporal correlations.” In proceedings of the*IEEE International Conference on Computer Aided Design*, pages 294-299, 1994.

[12] F. Najm, R. Burch, P. Yang, I. Hajj. “Probabilistic simulation for reliability analysis of CMOS VLSI circuits.” In *IEEE Transactions on CAD*, 1990, volume 9, pages 439-450.

[13] F. N. Najm. " Transition density: A new measure of activity in digital circuits. "*IEEE Transactions on Computer-Aided Design of Integrated Circuits and Systems*, 12(2):310-323, February 1993.

[14] M. Pedram, B. T. Preas. “Interconnection length estimation for optimized standard cell layouts.” In proceedings of the *IEEE International Conference on Computer Aided Design*, pages 390-393, November 1989.

[15] R. A. Powers. " Batteries for low power electronics. " *Proceedings of IEEE*, 38(4):687-693, April 1995.

[16] S. Rajgopal and G. Mehta. “Experiences with simulation-based schematic level current estimation.” In proceedings of the *International Workshop on Low Power Design*, pages 9–14, April 1994.

[17] C. Small, “Shrinking devices put the squeeze on system packaging.”*EDN*, vol. 39, no. 4, pages 41-46, Feb. 17, 1994.

[18] C-Y. Tsui, M. Pedram, and A. M. Despain. “Efficient estimation of dynamic power dissipation under a real delay model.” In proceedings of the *IEEE International Conference on Computer Aided Design*, pages 224–228, November 1993.

[19] S. Turgis, N. Azemard and D. Auvergne. " Explicit evaluation of short circuit power dissipation for CMOS logic structures. " In *Proceedings of the 1995 International Symposium on Low Power Design*, pages 129-134, April 1995.

[20] A. Tyagi. "Hercules: A power analyzer of MOS VLSI circuits" In Proceedings of the*IEEE International Conference on Computer Aided Design*, pages 530–533, November 1987.

[21] H. J. M. Veendrick. " Short-circuit dissipation of static CMOS circuitry and its impact on the design of buffer circuits. " *IEEE Journal of Solid State Circuits*, 19:468–473, August 1984.

CHAPTER 2

Technology Independent Power Analysis and Modeling

This chapter presents a study on the relation between the power consumption of a Boolean network before and after technology mapping, under a zero delay model. Using a new signal tracing technique, it is shown that the structure of a network before mapping directly impacts the power consumption of the mapped network. This analysis will demonstrate that the contribution of each node in the unmapped network to the load and switching activity of the mapped network can accurately be estimated. Having provided a technique that provides an accurate power estimate on a node by node basis, a strong case is made for technology independent power optimization. Two load models for nodes in the technology independent network are also presented. These models will be used to minimize power consumption during technology independent power optimization procedures presented in the following chapters.

2.1 Power Modeling Overview

Dynamic power accounts for a significant part of the power consumption in digital CMOS circuits. The average dynamic power consumption of a CMOS gate n_i in a synchronous CMOS circuit given by equation 1.1, can accurately be used to measure the power consumption of a network after it is mapped to the gates in the target technology. However, the power consumption of a Boolean network, before mapping, cannot accurately be measured using this equation. The reason is that the relationship between the power consumption of nodes in a technology independent network, and the power consumption of nodes in the technology mapped network is not clearly understood. This means that an estimate obtained by using the above equation during technology independent phase of optimization might lead to inferior results, since the load, switching activity, and the number of nodes in the mapped network might be completely different from the corresponding values in the technology independent network.

Moreover, an effective technology independent power optimization technique requires an accurate estimate of the impact of each node in the technology independent network to the power consumption of the mapped network. This will allow for guiding the optimization procedure by selecting node implementations that result in a minimal power solution after the network is mapped. In other words, such an estimate will help in minimizing the power consumption of the mapped network by performing local optimization during the technology independent phase of optimization.

In order to develop such a model, the transformations that result in the final technology mapped circuit need to be clearly understood. Logic synthesis consists of technology independent logic optimization operations, technology mapping, and post mapping optimizations. The immediate goal for technology mapping and post-mapping optimizations is to minimize the switched capacitance, plus the internal power consumption of gates in the netlist. As stated before, the same cost cannot, however, be applied during technology independent optimization. The goal of this chapter is to establish an immediate goal (i.e. a cost defined in terms of network parameters before mapping) for technology independent logic optimization. This immediate goal is then used for the technology independent optimization procedures.

The technology independent cost model is dictated by the technology mapping procedure because this is the operation that transforms the Boolean network into a netlist of gates. Therefore, a technology independent power cost model should be created, while considering the technology mapping technique that is used. Boolean matching [5] or structural mapping techniques [1],[3],[9] can be used to perform technology mapping. The structural tree-based mapping algorithm is a commonly used technique, and is used in many logic synthesis tools including our synthesis environment [2]. This algorithm partitions the unmapped Boolean network into a collection of trees where each node in this tree is a basic primitive node (i.e. 2-input NAND gate or Inverter). Gates in the target technology are also represented using tree structures consisting of the same primitive nodes. The process of technology mapping proceeds by optimally mapping trees in the Boolean network to the sub-trees (tree primitives) representing the library gates using tree matching techniques. The low power technology mapping technique developed within our power optimization system (described in chapter 7) uses the fundamental ideas behind structural technology mapping.

This chapter shows that assuming a structural technology mapping approach based on tree matching techniques, the goal for technology independent logic synthesis should be to generate a maximally decomposed network [10] where the product of zero-delay model switching activity of each node output, and its load in the factored form (as defined in section 2.2.3), is minimized. It is shown in this chapter that assuming a structural mapping technique, this cost function, directly minimizes a large part of the power consumption in the technology dependent network. Other sources of power consumption, such as power due to spurious activity and also internal gate power consumption, are ignored in this model. These sources of power, however, can more effectively be minimized during the technology mapping and post-mapping optimization steps.

Technology independent logic synthesis is the process of minimizing the user defined network cost by taking advantage of common sub-expressions within and between node functions [6], and the flexibility provided by don't care conditions derived from network structure, and provided by the user [7]. In this context, the network is optimized by introducing new nodes in the network, removing nodes that do not reduce the network cost, and also changing node functions by using don't care conditions. The process of technology independent logic optimization can, therefore, be described as a sequence of operations where new nodes with low switching activity are inserted into the network, nodes with high activity are removed from the network, while redistributing the load from high activity nodes to low activity nodes. As stated before, it is critical that network area is not drastically increased because this increase can potentially offset the power reduction.

This chapter first presents a study on the relationship between the switching activity and load of a network before and after technology mapping. This analysis is performed using a procedure called *"equivalent activity signal tracing"* or, in short, *"signal tracing"*. This analysis will be used to identify all nodes n in the unmapped network for which there exists one or more nodes m_i (gates) in the mapped network such that either n is derived by inverting or buffering m_i, or vice versa. Note that nodes n and m_i will have the same switching activity under a zero-delay model. The result of this tracing will help in identifying the nodes that will directly impact the power of mapped network. This means that any optimization done on these nodes will directly impact the power of the mapped network. The second part studies how the load at the output of nodes change after the mapping procedure. It is shown that the load on these nodes can accurately be estimated using our load estimation techniques.

The remainder of this chapter is organized as follows. Section 2.2 presents definitions and background. Section 2.3 discusses signal tracing and makes observations based on the results of the experiments.

2.2 Preliminaries and Definitions

2.2.1 Computing switching activities

The following method is consistently used to compute the switching activity for all nodes in the network for purposes of power estimation and optimization. Signal probabilities at the primary inputs are used to compute the signal probability for each internal node by building its global BDD. The switching activity of the node is then computed using its signal probability, assuming temporal independence at the primary inputs. Given the signal probability sp_n for an internal node *n*, the switching activity for *n* is computed as $2*sp_n*(1-sp_n)$. The same method is used consistently for estimates, before and after technology mapping.

The following discussions regarding power estimation and optimization will be valid, independent of the method used for computing the switching activities at each node. The methods discussed here make use of the switching activity information to make estimates, or guide the optimization process. Given a different zero-delay model technique to compute switching activities (for example, one that accounts for temporal correlations), these methods will be able to adjust to the new network parameters.

2.2.2 Power consumption vs. power contribution

Node power consumption is defined as the power consumed at the output of a node. This power is related to the output load and switching activity. The network power consumption can be obtained by summing the node power for all nodes in the network. The power contribution of a node *n*, however, reflects the part of the network power that is consumed due to the node being present in the network. Thus the power contribution of a node is the power consumed at the output of the node, plus the power consumed on the edges that fan into the node. Node power is used to compute the total network power consumption, while power contribution of a node is used to study the feasibility of a node being in the network.

2.2.3 Load estimation

Previous methods for power estimation have used the number of fanouts for a node as an estimate of the load at the output of a node [8]. A more accurate load model uses the "load in the factored form" to compute the load at the output of a node. Given $FL(i, k)$, the number of times variable i is used in the factored form function of node k, equation (2.1) is used to compute the load at the output of node n_i:

$$L_i = \sum_{k \in fanouts} FL(i, k) \tag{2.1}$$

The following example shown in figure 2.1 illustrates this computation.

Example 2.1:

In order to use the new method we first observe the following:

$FL(4,1) = 1$ $\quad FL(5,1) = 2$

$FL(6,1) = 1$ $\quad FL(1,2) = 3$

$FL(1,3) = 2$

Then

$L_1 = 5$ $\quad L_5 = 2$

$L_4 = 1$ $\quad L_6 = 1$

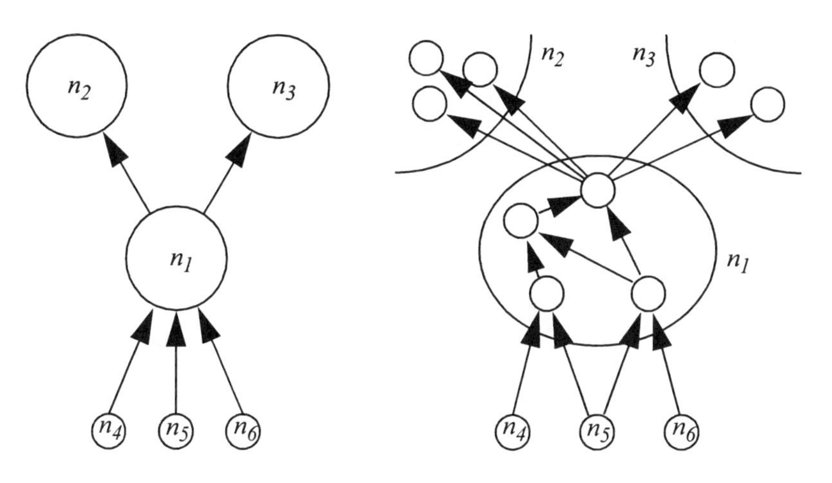

Figure 2.1 Load in the factored form

2.3 Signal Tracing for Power Analysis

In order to perform effective technology independent power optimization, an accurate estimate of the contribution of each node to the power consumption of the mapped network is needed. The goal of the signal tracing analysis presented here is to provide a basic understanding of how the power consumption of a network after mapping relates to the power consumption of each node before mapping. The result of this analysis is then used to draw conclusions on what approach should be used during the technology independent power optimization.

A "*signal*" S_n is defined as the global function on the output of a node *n*. Equivalent signals are defined as follows. The signals on all nodes that are generated by inverting or buffering node *n*, are equivalent to signal S_n. Also, signals on the output of nodes *m* that are used to generate *n* by inverting or complementing *m* are also equivalent to S_n. Nodes after mapping that are equivalent to a node *n* in the technology independent network are called "*born-again nodes.*" Note that the output of born-again nodes and node *n* will have the same switching activity assuming a zero-delay model. In the context of this chapter, node equivalence and signal equivalence are used interchangeably. Figure 2.2 shows part of a Boolean network before and after technology decomposition and inserting branch point inverters. In this example, signals m_1, m_2, m_3, m_4, m_5, m_6 are all equivalent to signal n_1 in the original network. Note that after technology mapping, some signals will be covered by complex gates.

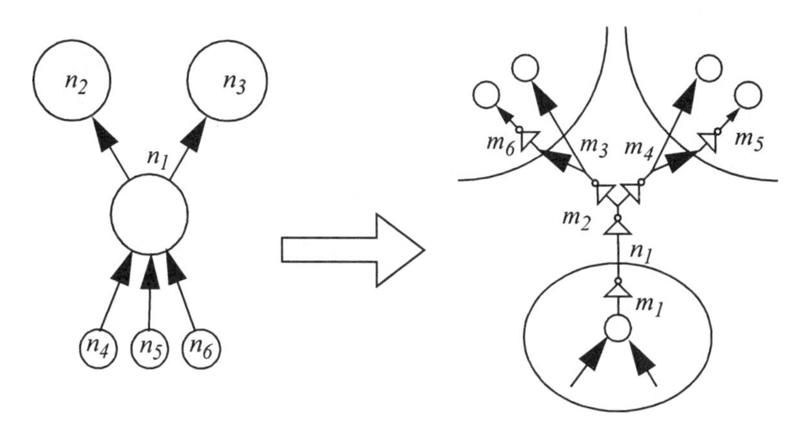

Figure 2.2 Equivalent signals

The analysis in this section will show the dependence of the power of a mapped network to its structure before mapping. This analysis will also help in identifying node properties that contribute to lower power after mapping, and will help us understand the impact of a node before mapping to the power of the mapped network.

The following procedure is used to trace the signals in a network. Before the network is mapped, all signals that will be traced are marked by building a table for each node containing all nodes equivalent to that node. At this point, the equivalence table for a node n will only contain node n if no inverters exist in the network. After this stage, every time a new node is being placed in the network a check is made to see if it is equivalent to any signals initially marked. For example, during double inverter insertion, both inverters are placed in the equivalence table of the node that drives the inverters. Also, during the mapping procedure, every time a node is removed from the network by covering it with a larger node, that node is removed from the equivalence table of any node that it might belong to. Note that at this point, a table might become empty. This means that the original signal marked does not have any equivalent signals in the mapped network.

All nodes in a mapped network can be analyzed by studying the equivalence tables. All nodes that are not in any of the equivalence tables have introduced new signals into the network. The nodes generating these signals are called "*new*" nodes. All other signals are now in one and only one equivalence table T_i. This means they are equivalent to the original signal S_i marked in the unmapped network. Note that more than one signal in the mapped network might be equivalent to a marked signal before mapping. This is the case when multiple copies of the signal and its complement have been used in the mapped network because of inserting multiple inverters for the same signal. "*Born-again*" nodes are now all nodes in the mapped network that are in an equivalence table. "*Lost signals*" are the ones that were marked initially, but have no equivalent signal in the mapped network. The nodes generating lost signals are called "*lost nodes*". "*New nodes*" are nodes in the mapped network that are not in any equivalence table.

The following experiments first analyze what signals in an unmapped network have equivalent signals in the mapped network. A second experiment analyzes how the load for these signals change as the network is mapped. Conclusions are then made, based on our observations in these experiments. The following experiments use the standard benchmark circuits from the Microelectronics Center of North Carolina [4].

2.3.1 Born-again nodes

Table 2.1 shows the results of signal tracing over the set of shown examples. The results here are intended to show the number of new and marked nodes in the mapped network. Marked nodes are the nodes that were marked for tracing before mapping. As the results show, the signals initially marked, almost always, have an equivalent signal in the mapped network. The reason for this is that signal tracing process was performed after selective collapse, and the extraction procedures have been performed. This means that all nodes in the network have more than one fanout, and if they have one fanout, the single fanout is used more than once in their fanout node. Even if selective collapse was not performed after extraction, the same trend in results is expected since extraction only extracts nodes that save area, and hence have more than one fanout. This also means that after node decomposition all nodes that were initially marked now have multiple fanout. Tree matching based technology mappers usually cross the multiple fanout only when they can take advantage of the branch point inverters. However, the signal on these inverters is equivalent to the signal that has multiple fanout and was marked initially. When the mapping is done, the originally mapped node will still have an equivalent node, whether in true or complemented form.

The results in this table show that almost all signals in the technology independent network have an equivalent signal in the unmapped network. Also note that the switching activity on equivalent signals is the same under a zero-delay assumption. This means that switching activity of born-again nodes in the mapped circuit is directly proportional to the switching activity of the signals in the unmapped network. This also means that minimizing the switching activity of signals in the unmapped circuit will minimize the switching activity of the born-again nodes. The analysis in the next section shows that the born-again nodes usually have a much bigger load than new nodes. This means that even though the number of born-again and new nodes seem to be in the same range, minimizing the switching activity of the born again nodes will have a more significant impact on the power consumption of the mapped network.

In order to study the effect of mapping on the networks derived from the same network, the following experiment is performed. A Boolean network is optimized for area using 40 different scripts. The scripts were generated using different options for extraction, simplification, and decomposition during the technology independent optimization. This generated a set of examples that are used as the subject examples.

2.3.2 Tracing the load on signals

The load estimation technique presented in section 2.2.3 is used to estimate the node loads before technology mapping. The load of a node after mapping is computed as the sum of the load for all its equivalent nodes after mapping. Two sets of results are presented. First, the load of node n before mapping is compared to the number of fanouts for all nodes equivalent to n after mapping. The load of node n before mapping is then compared to the sum of the load given by the library on all nodes equivalent to node n.

Rather than reporting the absolute value of the load on each node in each network, the following format is used to report these results. "*load difference*" is defined for a node n to be the difference between load after mapping and load estimate for n before mapping. A table is then presented that lists for each possible load difference, the number of nodes in all networks which had that load difference.

Table 2.2 gives the result of load tracing. In this table our load estimate for signal n before mapping is given as the number of fanout in the factored form of n. The load after mapping is given as the sum of the number of fanouts for all signals equivalent to n after mapping. Table 2.3 gives the result of load tracing when the load after mapping is derived from the library parameters. Lsi10k library is used for these experiments. The column 2 in each table gives the load difference for all nodes analyzed. Column 3 excludes the primary output nodes. Note that primary output nodes in this case refer to direct fanins of the nodes that have a type primary output, so it is possible for these nodes to be used in implementing other nodes in the network. Column 4 includes only internal nodes, and column 5 has only primary input nodes. Note that column 1 should be considered to make deductions on the results since all nodes included in this column can potentially change their load after mapping. The analysis excludes primary outputs since neither zero delay switching activity nor the load on these nodes can be modified during the technology independent optimization.

As is evident from the results in table 2.2, almost all the time, the actual number of fanouts after mapping is either the same, or one more than our estimate before mapping. The reason for cases where the fanout is only one more is when a signal in the network after mapping implements the complement of the original signal marked for tracing. This additional load for an inverter is the reason for a larger number of load estimates being off by 1 fanout. The results in table 2.3 are expected since the loads in the library are normalized to 1. The reason that the majority of the nodes in this case are off by 1.5 is that the inverters used to implement the complement of the function have a fanin load of 1.5 in most cases.

The previous section showed that almost all the nodes present in the network before technology mapping are present in the mapped network in the form of one or more equivalent signals. The results in this section show that for all such signals, the number of fanout, and hence, the load for the equivalent signals in the mapped network can accurately be estimated. This means that the impact of a node to the power consumption of the mapped network can also be accurately measured. The next section shows that the power on born-again nodes accounts for a significant part of the power consumption of the mapped network. Also, past analysis has known that the power on born-again nodes can directly be reduced by reducing the power on all visible signals in the unmapped network. This means that the primary objective of the technology independent power optimization should be to minimize the product of switching activity, times the number of fanouts in the factored form for all visible signals in the network, before technology mapping.

The results in this section also show that measuring load in the factored form is more accurate than estimating load as the number of fanouts for the node. The reason is that we know the number of load in the factored form of a node is always more than the number of fanouts for the same node. The results of this section also show that the actual load on a signal after mapping is almost always the same, or one more than the number of fanout in the factored from before mapping. The conclusion can then be made that any estimate obtained by using the number of fanouts, as the load will always be less accurate than using the number of fanout in the factored from of the node.

An added advantage of this accuracy is that wiring models which directly depend on the number of fanouts of a node after mapping can also be applied during the technology independent phase of power optimization.

2.3.3 Power analysis using tracing information

This section shows the contribution of new nodes and born-again nodes to the power consumption of the mapped network. All powers reported after mapping are derived from the library load parameters. *"Old power"* corresponds to summing the product of switching activity times the load factored form over all nodes that are being traced (all nodes in the network after mapping).

Table 2.5 shows the results of trace analysis for power. Column 2 gives the power computed before mapping on all signals to be traced. Column 3 gives the power before mapping on primary inputs. Column 4 gives the power on the born-again nodes after mapping, and column 5 gives the power on primary inputs after mapping. Column 6 gives the power of nodes before mapping that have no equivalent node after mapping. Column 7 gives the power on the signals created after mapping, or the new power. Column 8 gives the ratio of the new power to the born-again power. The average of this ratio for all examples is 14% and ranges from 0% to 43%. The standard deviation for these ratios is 0.09.

2.3.4 Observations

A number of conclusions are made from the results of the trace analysis. It is evident that almost all of the signals that have a factored load of 2 or more will be present in the mapped network. It is also shown that the load in the factored form is an accurate measure of the load of the network. It is shown that the number of fanouts for a born-again signal after mapping is almost always the same or one more than the number of load in the factored form before mapping. The actual load after mapping is almost always the same or 1.5 units more than the load predicted before mapping (using lsi10k library).

The important conclusion derived from these observations is that any power optimization performed on visible signals during the extraction procedure will directly impact the power of the mapped network. The load estimates are also shown to be close to the loads after mapping.

2.4 Power Models for Node Functions

Given a node n_i and fanin node n_j, *factored load* of n_j with respect to n_i, $FL(n_j, n_i)$ is defined as the number of times variable n_j is used (in positive or negative form) in the factored form expression of node n_i. The *cube load* of n_j with respect to n_i, $CL((n_j, n_i)$ is also defined as the number of times variable n_j is used (in positive or negative form) in the sum-of-products form of node n_i. For the case shown in 2.3, $FL(n_5,n_1)=2$ while $CL(n_5,n_1)=3$.

Using these definitions, two power cost functions for a node are defined. Note that the cost functions presented here represent the contribution of a node to the power consumption of the network since they include both the power at the input and the output of the node. This means that these costs cannot directly be used to find the power cost of the network since the power on the output of all internal and primary input nodes will be counted twice.

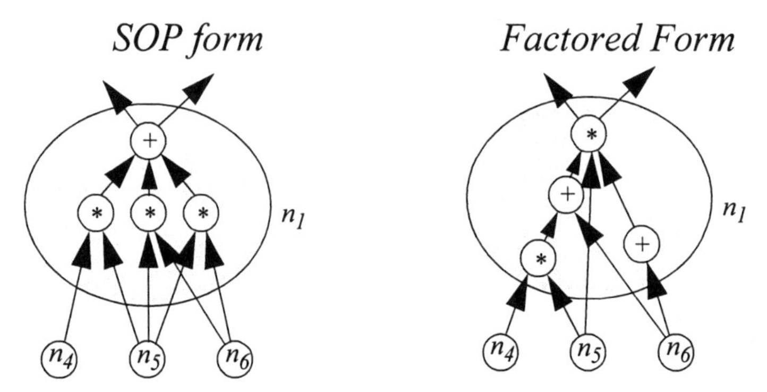

Figure 2.3 Node power modeling

The sum-of-products cost function gives the power consumed by the node if the node is implemented in a sum-of-products form. Given a node n_i with output function f and cubes $(c_1,.. c_N)$, the power cost of n_i in the sum of products form is computed as:

$$P^{SOP}_{n_i} = sw(f) \cdot \sum_{n_k \in fanouts(n_i)} CL(n_i, n_k) + \sum_{c_m \in cubes(n_i)} sw(c_m) + \sum_{n_j \in fanins(n_i)} sw(n_j) \cdot CL(n_j, n_i) \tag{2.2}$$

The first term in this equation corresponds to the power consumption at the output of the node. The second term corresponds to the power consumption at the output of the gates implementing each product term of the function, and the last term accounts for the power on the inputs to the node.

Given a node n_i, the power cost of the node in the factored form is computed as:

$$P^{FAC}_{n_i} = sw(f) \cdot \sum_{n_k \in fanouts(n_i)} FL(n_i, n_k) + \sum_{n_m \in Internal(n_i)} sw(n_m) + \sum_{n_j \in fanins(n_i)} sw(n_j) \cdot FL(n_j, n_i) \tag{2.3}$$

where $Internal(n_i)$ is the set of all internal nodes in the factored form representation of n_i.

Note that is equations 2.2 and 2.3, the load for internal nodes is assumed to be 1. This is a valid assumption since for SOP representation, each node implementing a product term only fans out to one node. The procedure for fully factorizing a node [10] and its extension to consider power (see chapter 5) also guarantee that each internal node has only one fanout node.

2.5 Technology Independent Power Model

This chapter introduced a signal tracing approach that allowed us to track what nodes in the unmapped network contribute to a network's load, switching activity, and power after mapping. This analysis showed that almost always all nodes that have a factored load of more than 1 before technology mapping will have equivalent signals after mapping. It was also shown that for a node n before mapping, the load due to n after mapping can accurately be estimated. The results showed that before technology mapping, all nodes with a fanout of more than 1 will directly impact the power of the mapped network in the form of born-again nodes, and the born-again power can accurately be estimated. It was also shown that the majority of a network's power after mapping is due to born-again nodes.

The results of signal tracing also shows that load in the factored form is always a better estimate of the load before mapping. The trace analysis showed that load in the factored form is almost always less than, or equal to, the actual number of fanouts after mapping. At the same time, load in the factored form is always the same or more than the number of fanouts for a node. This means that load in the factored form will always be a better estimate of a nodes load than the number of fanout.

Since the born-again power dominates for most circuits, the goal of a technology independent optimization procedure can be redefined to minimize the power on the born-again nodes after mapping. An increase in the new power of the mapped network can be tolerated if this increase in power is more than off-set by a reduction in the born-again power. New power can also effectively be minimized during technology mapping. Therefore in order to minimize the born-again power, the product of switching activity times load in the factored form of all nodes in the technology independent network, need to be minimized.

example	pre-map		post-map				new/ba
	old power	pi old power	ba power	pi ba power	lost power	new power	
9symml	37.0	13.5	54.3	20.2	0.0	2.0	0.04
C1355	176.5	104.0	234.4	128.0	0.0	81.9	0.35
C1908	187.5	87.5	244.4	111.2	0.0	33.0	0.13
C432	120.6	58.5	230.3	113.0	0.0	99.8	0.43
C499	201.5	104.0	279.7	128.0	0.0	41.6	0.15
C5315	838.1	372.0	1082.2	518.8	36.8	149.3	0.14
C880	180.2	118.5	245.9	152.8	0.0	24.4	0.10
alu2	161.4	77.5	190.0	82.5	0.0	39.6	0.21
alu4	389.1	204.0	452.9	217.5	0.0	87.0	0.19
apex6	344.9	250.5	479.7	360.8	0.0	64.1	0.13
apex7	122.0	90.0	158.9	119.0	0.0	17.5	0.11
b9	54.4	44.0	77.7	62.2	0.0	8.3	0.11
c8	70.7	55.5	90.5	71.2	0.0	14.7	0.16
cc	28.9	23.5	41.4	33.8	0.0	9.1	0.22
cht	73.4	46.5	109.4	81.8	0.0	1.1	0.01
cm150a	23.5	23.5	28.0	28.0	0.0	7.4	0.26
cm162a	17.4	13.0	25.6	19.8	0.0	2.5	0.10
cm163a	17.5	12.5	25.7	18.5	0.0	2.6	0.10
cm42a	15.1	9.5	18.1	12.5	0.0	0.9	0.05
cm85a	20.6	18.5	27.4	25.2	0.0	3.8	0.14
cmb	24.0	24.0	34.0	34.0	0.0	5.3	0.16
comp	75.1	48.0	108.9	72.0	0.0	9.0	0.08
count	52.0	49.5	80.2	76.0	0.0	21.5	0.27
cu	26.2	24.5	32.5	30.5	0.0	4.1	0.13
decod	19.0	8.0	24.1	11.8	0.0	0.0	0.00
des	1395.4	522.5	1711.8	707.2	0.0	157.0	0.09
example2	142.1	114.5	184.0	149.2	0.0	11.9	0.06
f51m	60.8	40.0	75.3	47.0	0.0	13.9	0.18
frg1	59.5	49.5	80.8	67.8	1.0	20.2	0.25
frg2	352.8	238.0	459.5	321.8	0.0	104.3	0.23
k2	229.3	151.0	274.8	186.5	0.0	20.3	0.07
lal	39.1	33.0	61.3	51.5	0.0	6.0	0.10
ldd	33.1	18.5	43.4	25.2	0.0	4.5	0.10
mux	39.4	31.5	24.5	18.5	0.0	6.5	0.26
my_adder	88.0	65.5	147.8	104.8	0.0	37.6	0.25
pair	686.8	477.5	850.2	595.5	0.0	134.9	0.16
parity	25.0	16.0	43.0	28.0	0.0	9.0	0.21
pcle	26.4	22.0	41.1	35.5	0.0	3.5	0.09
pcler8	33.4	26.0	49.3	34.2	0.0	3.6	0.07
pm1	25.7	21.0	38.6	31.0	0.0	1.4	0.04
rot	307.3	207.5	417.8	282.8	0.0	50.6	0.12
sct	30.1	23.5	46.4	36.8	0.0	3.5	0.08
tcon	16.0	16.0	28.8	28.8	0.0	0.0	0.00
term1	123.4	83.0	164.6	109.0	0.0	30.7	0.19
too_large	163.5	101.0	212.4	127.2	0.0	35.2	0.17
ttt2	108.3	88.5	127.4	101.5	0.0	31.3	0.25
unreg	41.8	34.0	67.1	58.0	0.0	7.6	0.11
vda	126.6	55.0	147.8	70.2	0.0	14.1	0.10
x1	132.4	109.5	185.1	153.2	0.0	28.0	0.15
x2	25.0	20.5	32.9	26.2	0.0	3.5	0.11
x3	356.9	261.0	494.4	368.2	0.1	75.9	0.15
x4	172.3	126.0	224.7	167.0	0.0	29.7	0.13
z4ml	18.0	10.5	28.2	15.8	0.0	0.0	0.00

Table 2.4 Effect of technology mapping on born-again power

example	marked nodes	lost nodes	new nodes	example	marked nodes	lost nodes	new nodes
9symml	27	0	5	f51m	43	0	34
C1355	155	0	154	frg1	44	1	44
C1908	180	1	63	frg2	462	0	219
C432	85	0	251	k2	388	0	160
C499	179	0	112	lal	69	0	15
C5315	618	26	277	ldd	65	0	16
C880	172	0	66	mux	27	0	18
alu2	74	0	134	my_adder	82	0	47
alu4	158	0	335	pair	637	0	363
apex6	394	0	174	parity	27	0	9
apex7	149	0	53	pcle	44	0	7
b9	87	0	25	pcler8	68	0	15
c8	74	0	46	pm1	43	0	5
cc	56	0	17	rot	406	0	140
cht	128	0	3	sct	54	0	10
cm150a	23	0	15	tcon	41	0	0
cm162a	30	0	7	term1	87	0	87
cm163a	31	0	7	too_large	98	0	129
cm42a	28	0	4	ttt2	87	0	74
cm85a	20	0	14	unreg	71	0	31
cmb	25	0	14	vda	215	0	75
comp	61	0	24	x1	131	0	82
count	81	0	48	x2	27	0	13
cu	37	0	11	x3	390	2	158
decod	47	0	0	x4	260	0	78
des	1142	0	789	z4ml	20	0	0
example2	247	0	40				

Table 2.1 Node trace analysis during technology mapping

load difference	all nodes	no po	no po no pi	only pi
-17	1	1	0	1
-16	0	0	0	0
-15	0	0	0	0
-14	0	0	0	0
-13	1	1	0	1
-12	0	0	0	0
-11	0	0	0	0
-10	2	2	0	2
-9	0	0	0	0
-8	1	1	1	0
-7	2	2	0	2
-6	0	0	0	0
-5	4	4	2	2
-4	0	0	0	0
-3	7	7	2	5
-2	16	16	5	11
-1	77	77	25	52
0	5062	1864	1356	508
1	2822	2822	1081	1741
2	138	138	68	70
3	25	25	14	11
4	4	4	3	1
5	5	5	3	2
6	11	11	4	7
7	0	0	0	0
8	0	0	0	0
9	1	1	0	1
10	0	0	0	0
11	0	0	0	0
12	3	3	1	2
13	4	4	3	1
14	0	0	0	0
15	6	6	4	2
16	2	2	1	1
17	0	0	0	0

Table 2.2 Effect of technology mapping on load values

load difference	all nodes	no po	no po no pi	only pi
-17	1	1	0	1
-16	0	0	0	0
-16.5	1	1	0	1
-12.5	1	1	0	1
-9.5	2	2	0	2
-5.0	1	1	1	0
-4.5	1	1	0	1
-4.0	1	1	0	1
-3.5	2	2	1	1
-3.0	4	4	2	2
-2.5	4	4	0	4
-2.0	10	10	5	5
-1.5	5	5	0	5
-1.0	35	35	17	18
-0.5	27	27	0	27
0.0	4789	1591	1217	374
0.5	102	102	18	84
1.0	128	128	114	14
1.5	1905	1905	309	1596
2.0	712	712	679	33
2.5	223	223	38	185
3.0	118	118	113	5
3.5	46	46	7	39
4.0	22	22	20	2
4.5	4	4	0	4
5.0	9	9	9	0
5.5	3	3	0	3
6.0	6	6	4	2
6.5	2	2	0	2
7.0	6	6	5	1
8.0	2	2	2	0
9.5	1	1	0	1
12.0	1	1	1	0
13.0	2	2	0	2
14.0	3	3	3	0
15.0	1	1	1	0
15.5	1	1	0	1
16.0	6	6	4	2
16.5	1	1	0	1
17.0	2	2	2	0
17.5	1	1	0	1
20.0	1	1	0	1
26.0	3	3	1	2

Table 2.3 Effect of technology mapping on library load values

example	old power	pi old power	ba power	pi ba power	lost power	new power	new/ba
sc111	945.9	514.0	1212.2	605.5	0.0	139.5	0.12
sc112	951.7	514.5	1215.7	606.0	0.0	139.3	0.11
sc121	945.9	514.0	1212.2	605.5	0.0	139.5	0.12
sc122	955.4	518.0	1219.2	611.8	0.0	140.1	0.11
sc131	945.9	514.0	1212.2	605.5	0.0	139.5	0.12
sc132	955.4	518.0	1219.2	611.8	0.0	140.1	0.11
sc141	883.9	380.0	1209.6	536.0	30.0	113.0	0.09
sc142	936.7	489.0	1200.4	591.2	0.0	134.6	0.11
sc211	919.2	498.5	1151.0	604.2	8.0	168.4	0.15
sc212	935.4	500.5	1171.5	607.0	4.6	171.6	0.15
sc221	919.2	498.5	1151.0	604.2	8.0	168.6	0.15
sc222	947.7	519.5	1191.4	633.5	4.1	180.6	0.15
sc231	919.2	498.5	1151.0	604.2	8.0	168.6	0.15
sc232	947.7	519.5	1191.4	633.5	4.1	180.6	0.15
sc241	867.1	375.5	1159.5	517.0	39.2	136.4	0.12
sc242	940.7	507.0	1178.2	601.2	2.1	177.5	0.15
sc311	922.0	506.0	1176.5	613.0	2.9	156.5	0.13
sc312	936.2	506.5	1177.9	613.5	2.9	162.0	0.14
sc321	922.0	506.0	1176.5	613.0	2.9	156.5	0.13
sc322	942.7	517.0	1191.9	628.5	5.1	172.3	0.14
sc331	922.0	506.0	1176.5	613.0	2.9	156.5	0.13
sc332	942.7	517.0	1191.9	628.5	5.1	172.3	0.14
sc341	878.7	377.0	1175.4	529.8	36.2	127.0	0.11
sc342	965.0	516.5	1222.5	623.5	3.7	158.5	0.13
sc411	837.8	365.0	1101.2	515.8	8.2	142.2	0.13
sc412	931.2	498.5	1134.4	589.2	2.4	232.9	0.21
sc421	944.6	510.0	1206.7	625.2	7.7	160.8	0.13
sc422	1004.3	540.0	1273.6	659.5	4.7	176.9	0.14
sc431	929.7	500.0	1147.1	604.0	6.8	159.4	0.14
sc432	997.1	536.0	1227.7	648.0	3.0	176.3	0.14
sc441	839.6	372.5	1102.8	523.5	4.8	136.4	0.12
sc442	924.8	490.5	1129.0	579.8	1.4	231.1	0.20
sc511	837.8	365.0	1101.2	515.8	8.2	142.2	0.13
sc512	931.2	498.5	1134.4	589.2	2.4	232.9	0.21
sc521	944.6	510.0	1206.7	625.2	7.7	160.8	0.13
sc522	1004.3	540.0	1273.6	659.5	4.7	176.9	0.14
sc531	929.7	500.0	1147.1	604.0	6.8	159.4	0.14
sc532	997.1	536.0	1227.7	648.0	3.0	176.3	0.14
sc541	839.6	372.5	1102.8	523.5	4.8	136.4	0.12
sc542	924.8	490.5	1129.0	579.8	1.4	231.1	0.20

Table 2.5 Trace analysis for power of networks derived from C5315.blif

References

[1] K. Chaudhary and M. Pedram. "A near optimal algorithm for technology mapping minimizing area under delay constraints." In proceedings of the *ACM/IEEE Design Automation Conference*, June 1992.

[2] S. Iman and M. Pedram. "POSE: Power Estimation and Synthesis Environment," In proceedings of the *ACM/IEEE Design Automation Conference*, June 1996.

[3] K. Keutzer. "DAGON: Technology binding and local optimization by DAG matching." In proceedings of the *ACM/IEEE Design Automation Conference*, June 1987.

[4] R. Lisanke, editor. "FSM benchmark suite," Microelectronics Center of North Carolina, Research Triangle Park. North Carolina, 1987.

[5] F. Mailhot, G. DeMicheli. "Technology mapping using Boolean matching." In proceedings of European Conference on Design Automation, March 1990.

[6] R. Rudell. "Logic Synthesis for VLSI Design." Ph.D. thesis, University of California, Berkeley, 1989.

[7] H. Savoj. "Don't Cares in Multi-Level Network Optimization." PhD thesis, University of California, Berkeley, 1992.

[8] A. A. Shen, A. Ghosh, S. Devadas, and K. Keutzer. "On average power dissipation and random pattern testability of CMOS combinational logic networks." In proceedings of the *IEEE International Conference on Computer Aided Design*, November 1992.

[9] C-Y. Tsui, M. Pedram and A. M. Despain. "Power efficient technology decomposition and mapping under an extended power consumption model." In *IEEE Transactions on Computer Aided Design*, Vol.~13, No.~9 (1994), pages 1110--1122.

[10] A. Wang. "Algorithms for Multi-Level Logic Optimizations." Ph.D. Thesis, UC Berkeley, 1989.

Part II

Two-level Function Optimization for Low Power

CHAPTER 3

Two-Level Logic Minimization in CMOS Circuits

Boolean algebra is the primary mathematical tool used in designing the computing machines that have come into widespread use in the latter part of this century. Switching functions, described using two-valued Boolean algebra, are implemented using transistor circuits [4]. These transistors act as on-off switches, which is a natural implementation for representing the primitive operations in a Boolean algebra. Once the primitive operations of a Boolean algebra are implemented using transistor circuits, any Boolean function can be implemented using these primitive operations. Operations AND, OR and NOT form the primitive operations of Boolean algebra. Other variations of these primitive operations such and NAND and NOT can also be used as the set of primitive operations.

Two levels of logic are the minimum that is required to implement any Boolean function. The first level logic contains product terms (AND), and the second level logic contains sum terms (OR). It may also be necessary to use inverters to invert some of the inputs to the product terms. Using this representation, two-level operations are also called *Sum-Of-Products* (SOP) form representations. Two-level implementations of Boolean functions have thoroughly been studied, and techniques for optimum implementations of such structures have been proposed by early logic design researchers [6],[8]. Two-level implementations have remained in widespread use as the computer design techniques have matured over the past forty years. Introduction of PLAs [3] which implement two-level descriptions of Boolean functions, and later the use of standard cell techniques that allow the implementation of multi-level logic circuits, have resulted in continued interest in two-level logic design[1]. More advanced techniques have been proposed by researchers for both exact and approximate solutions for optimizing two-level representations of Boolean functions. ESPRESSO [1] presents a heuristic approach where novel techniques are used to efficiently produce good area solutions. ESPRESSO EXACT [9] presents an exact method for solving the minimum area solution.

The target cost function for two-level function optimization has traditionally been the number of product terms, followed by the number of product term inputs (literals). Minimizing this cost function is independent of the technology being used to implement the Boolean function. This means that the same area optimized implementation may be used for static and dynamic logic circuits. The cost functions used for measuring power, however, are different in static and dynamic circuits [2]. This means that different strategies need to be developed for different technologies.

Our goal in this chapter is to study the problem of optimizing a Boolean function in order to minimize power consumption in multi-level Boolean networks. Multi-level networks are usually implemented using static CMOS circuits where circuit implementation consists of single output nodes. Therefore, this chapter only considers power minimization for Boolean functions, assuming that the Boolean function is a single output function (meaning that all product terms of the function are only used in one OR term). Chapter 4 deals with PLA implementations of two-level Boolean functions.

Recent works on minimizing the power consumption of Boolean function have concentrated on heuristic approaches. A method is presented in [10] where the power cost function of the prime implicants was used to solve an ESPRESSO like procedure. This procedure did not, however, consider including non-prime implicants in the cover although these implicants can lead to more power reduction. Recently [12] presented a heuristic approach for minimizing the power consumption of a function. The major shortcoming of this approach is that it assumes the same signal probability at all inputs of the function which, in general, does not represent realistic conditions.

This chapter first presents an exact technique for minimizing the power consumption of the two-level implementation of a Boolean function implemented using static CMOS circuits. It then presents an analysis of the complexity of solving this problem when compared to solving the same problem for minimum area. Using this

1. Popular models used to represent multi-level circuits use two-level functions as their building blocks, and therefore two-level optimization techniques are a necessary component of multi-level design techniques.

analysis, a technique is proposed to limit the increased complexity of the problem by restricting the set of implicants which are being considered for inclusion in the final implementation.

3.1 Power Model for Two-Level Static CMOS Logic

Given a Boolean function *f* with product terms $Q=(q_1, ...q_M)$ and input set $V = (v_1,...v_N)$. The following model is used to estimate the power consumption of the two-level implementation of function *f* using static CMOS circuits.

The power consumption due to each product term q_i is given by:

$$Pow(q_i) = 0.5 \cdot \frac{V_{dd}^2}{T_{cycle}} \cdot \left((C_{AND} \cdot sw(q_i)) + \sum_{l_j \in lit(q_{i_i})} (C_{IN} \cdot sw(l_j)) \right) \tag{3.1}$$

The power consumption of the function is then given by:

$$Pow = 0.5 \cdot \frac{V_{dd}^2}{T_{cycle}} \cdot C_{OR} \cdot sw(f) + \sum_{q_i \in Q} Pow(q_i) \tag{3.2}$$

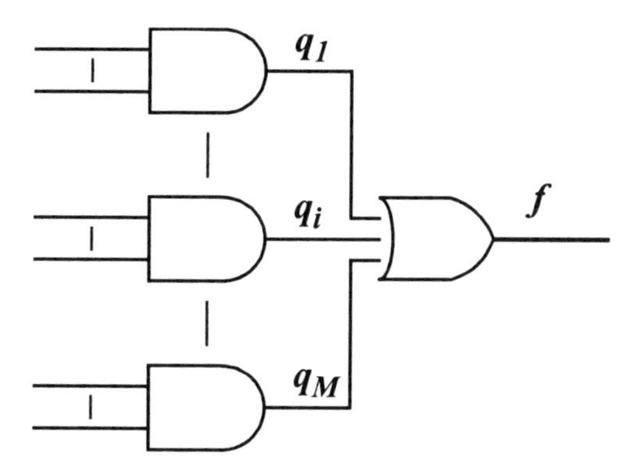

Figure 3.1 SOP form implementation of function f

where C_{OR} is the load seen by the output of the OR gate, C_{AND} is the load seen by the output of the AND gate and C_{IN} is the load seen by the inputs. *sw(i)* gives the switching activity of gate *i*. In the following we normalize this equation by setting

$$\frac{V_{dd}^2}{2 \cdot T_{cycle}} = 1$$

Note that the power consumption of any implementation of the circuit (NOR-NOR, NAND-NAND, etc.) can be estimated using equation 3.2 since a function *f* and its complement will exhibit the same switching activity when equation 3.3 is used.

3.2 Exact Minimization Algorithms

Exact methods cannot be applied to many large functions. Therefore, an exact optimization procedure cannot effectively be used in real circuits. However, an exact procedure is needed for two reasons. First, an exact solution can be used to measure the quality of heuristic approaches. Second, an exact procedure can be used to study the effectiveness of the optimization on the cost function which is being optimized. For example, during logic synthesis, area and power are closely related. This means that a low area solution tends to provide a low power solution as well. An exact method in this case provides insight into the maximum power savings that can be achieved as compared to a minimum area solution.

The problem of exact two level logic minimization for minimum area is stated as follows:

Problem: *Given a Boolean function f with input set $V = (v_1,...v_N)$, find a two-level implementation of the function f such that the number of product terms and then the number of literals in the sum-of-products form is minimum.*

Using the number of product terms and number of literals as the cost function, it is shown in [7] that the minimum cover of a Boolean function always consists of sum of prime implicants of the function. Therefore, this problem is solved by first generating the set of all prime implicants of the function [1]. The minimum area solution is then obtained by solving an exact minimum covering formulation of the problem where a subset of the primes are selected to cover the function.

The problem of exact two-level function minimization for low power is stated as follows:

Problem: *Given a Boolean function f with input set $V = (v_1, ... v_N)$ and signal probability sp(i) for each input, find a two-level implementation of the function f such that the power as given in equation 3.2 is minimum.*

The main difficulty in generating an exact minimum power solution is that compared to a minimum area solution which only requires prime implicants, other implicants may need to be considered while solving the covering problem. In the next section *Power Prime Implicants (PPIs)* are introduced and techniques for computing them are presented.

In what follows it is assumed that the value for $C_{AND} = C_{IN}$. This is a valid assumption since the functions being optimized are internal nodes of a Boolean network. Also the following notation will be used.

Given q_i, an implicant of function *f*, $q_i^{l_1, \ldots, l_k}$ represents the implicant generated by lowering literals $l_1, \ldots, l_k$ in implicant q_i. An *n-cube* is also defined as an implicant with *n* literals.

3.3 Power Prime Implicants

As stated in the previous section, only prime implicants need to be considered while minimizing a function for minimum area. This can be proved by noting that any solution that contains a non-prime implicant *q*, will be improved by making *q* prime and therefore reducing the number of literals (and therefore the area cost) in the final solution. This argument cannot, however, be applied when implicants are being considered for a minimum power solution. The following example illustrates this observation:

Example 3.1:

Assume $sp(a) = 0.9$, $sp(b) = sp(c) = 0.5$ and the following two-level implementations for function f:

$$F_1 = a \cdot b + b \cdot c$$

$$F_2 = a \cdot b + \bar{a} \cdot b \cdot c$$

Also assume $C_{OR} = C_{AND} = C_{IN} = 1$, Then:

$$Pow(F_1) = 3.05$$
$$Pow(F_2) = 2.90$$

This example shows that implementation F_2 provides a better power solution in spite of including a non-prime implicant. Even though the implementation for a non-prime implicant requires more literals and more transistors, overall, less power is consumed. This is because in static CMOS circuits a significant part of power is consumed when gate outputs change values. In this example, for implementation F_2, the reduction in power at the output of the second cube more than offsets the increase in the power due to including literal a, therefore reducing the overall power consumption. This example shows that area-power trade-offs can be made by including non-prime implicants in the set of candidate implicants when the covering problem is being set up. This observation motivates the definition for *Power Prime Implicants* of a function.

Definition 3.1*An implicant q_i is a Power Prime Implicant (PPI) if:*

$$\forall q_j: \qquad q_i \subset q_j \Leftrightarrow POW(q_i) < POW(q_j)$$

By definition, all prime implicants (PI) of a function are also power prime implicants (*PPI*). In what follows the set of all *PPIs* of the function will also include the set of all prime implicants of the function. Non-prime *PPIs* are the set of *PPIs* which are not prime implicants.

Definition 3.2*Predecessors cubes of an implicant q_i with n literals are defined as all implicants with n-1 literals obtained by raising a literal in q_i. The Successors cubes of q_i are defined as all implicants with n+1 literals obtained by lowering a literal in q_i.*

3.4 Generating the Set of all PPIs

A simple approach for generating all *PPIs* of a function *f* with *N* inputs is to generate all n-cubes of a function for $n=1,...,N$ and compare the power for each n-cube q with that of all other *PPIs* containing q. This approach, however quickly becomes intractable as the number of such cubes grows exponentially. An efficient algorithm for generating the set of all *PPIs* of a function can be developed by using the following lemmas.

Lemma 3.1 *Given an implicant q_i with $sp(q_i) = x$ and also assuming temporal independence at for at the inputs of the function, $POW(q^l_i) < POW(q_i)$ iff:*

$$sp_l \cdot (1 - sp_l) + x \cdot sp_l \cdot (1 - x \cdot sp_l) - x \cdot (1 - x) < 0 \tag{3.3}$$

or equivalently iff:

$$sp_l < \frac{x - x^2}{1 + x^2} \tag{3.4}$$

Proof: Assuming temporal independence at the inputs of the function [5]:

$$sw(l) = 2 \cdot sp(l) \cdot (1 - sp(l))$$

Also, assuming that POW_A represents the power consumption on the inputs of implicant q_i. Then:

$$POW(q_i^l) < POW(q_j) \Leftrightarrow POW(q_i^l) - POW(q_j) < 0$$

$$\Leftrightarrow POW_A + E(l) + E(q_i^l) - POW_A - E(q_j) < 0$$

Equation 3.3 is readily obtained from these two equations. Equation 3.4 is then obtained from Equation 3.3 by collecting terms and considering that all signal probability values are between zero and one.

■

Theorem 3.1 will show that for each *PPI* q of the function *(excluding prime implicants)*, there exists another *PPI* which is a predecessor cube of q. This result will then be used to significantly reduce the number of implicants of the function that need to be checked as a candidate *PPI*. Lemma 3.2 will be used to prove theorem 3.1.

Lemma 3.2 Consider implicant q_i with output signal probability x and implicant $q_{i,n}$ obtained by lowering n literals $l_1,\ldots,l_n$ in implicant q_i. Implicant $q_{i,n}$ is a *PPI* if at least one of $q_{i,1}$ is a *PPI* ($q_{i,1}$ represents an implicant obtained by lowering one of literals $l_1,\ldots,l_n$ in q_i).

Proof: The power increase after replacing p_i with $p_{i,n}$ is given by:

$$\sum_{i=1}^{n} sp_i \cdot (1 - sp_i) + x \cdot \prod_{i=1}^{n} sp_i \cdot \left(1 - x \cdot \prod_{i=1}^{n} sp_i\right) - x \cdot (1 - x) \tag{3.5}$$

We prove that if none of $q_{i,1}$ are *PPIs* (or equivalently $sp_i \geq (x - x^2)/(1 + x^2)$ for $i = 1, \ldots, n$), then equation 3.5 is always positive which means that $q_{i,n}$ is not a *PPI*. Therefore proving the claim of this lemma.

Let $T = (x - x^2)/(1 + x^2)$. Equation 3.5 is a convex equation in all variables sp_i.

This is shown by taking the partial derivative with respect to variable sp_i and showing that this partial derivative is always negative in the unit cube where all variables range from 0 to 1. This means that the minimum value of this equation is on the boundaries of its domain represented by $T \leq sp_i \leq 1$ for $i = 1, \ldots, n$. The boundary points for this domain are represented by sp_i values where each sp_i is either set to 1 or T. Therefore to prove this theorem, we need to show that equation 3.5 is positive for all p_i values where p_i takes a value of 1 or T. Assume k of the p_i values are set to T and *(n-k)* are set to 1. Then equation 3.5 simplifies to:

$$k \cdot T \cdot (1 - T) + x \cdot T^k \cdot (1 - x \cdot T^k) - x \cdot (1 - x) \tag{3.6}$$

Now we need to show that equation 3.6 is positive for $1 \leq k \leq n$ and $0.0 \leq x \leq 1.0$.

It is easy to show that $2 \cdot T \cdot (1 - T) \geq x \cdot (1 - x)$. This means that for *k>1*, equation 3.6 is always positive (using first and third terms). For *k=1*, equation 3.6 is equal to zero (reduces to lemma 3.1). This means that for all values of *k*, equation 3.6 is positive which in turn proves the claim of this lemma.

■

Theorem 3.1 *If a non-prime implicant q_i with n literals is a PPI then there exists $q_j \supset q_i$, such that q_j is a PPI and q_j has (n-1) literals.*

Proof: Consider implicant q_i with output signal probability x and implicant $q_{i,n}$ obtained by lowering n literals $l_1, \ldots l_n$ in implicant q_i. Based on lemma 3.2, if $q_{i,n}$ is a *PPI* then at least one $q_{i,1}$ (obtained by lowering one literal in q_i) is also be a *PPI*. The same proof can then be applied to $q_{i,1}$ and $q_{i,n}$ (also contained in $q_{i,1}$). This in turn means that at least one $q_{i,2}$ (obtained by lowering a literal in $q_{i,1}$) should also be a *PPI*. Recursively applying lemma 3.2 shows that at there is at least one $q_{i,(n-1)}$ containing $q_{i,n}$ which is also a *PPI* thereby proving the claim of this theorem.

■

This theorem greatly reduces the number of implicants that have to be checked as candidate *PPIs*. It also provides an efficient method for generating the set of implicants that need to be checked as candidate *PPIs*. Now, given an implicant q to be checked as a candidate *PPI*, the following theorem shows that it is not necessary to compare q with all *PPIs* containing q, thus providing an efficient method for checking q as a candidate *PPI*.

Theorem 3.2 *Given an implicant q_i with n literals and Q the set of all PPIs with n-1 literals and given that q_i has at least one predecessor cube in Q:*

$$q_i \text{ is a PPI } \Leftrightarrow \forall q_j \in Q, q_i \subset q_j, POW(q_i) < POW(q_j)$$

Proof: From theorem 3.1, we know that if an implicant q_i does not have a predecessor cube which is a *PPI* then q_i is not a *PPI*. The sufficiency for the statement follows from the definition of *PPIs*. We prove necessity as follows:

Assume q_i is not a *PPI*. This means that there exists at least one *PPI* q_k which contains q_i and $POW(q_k) < POW(q_i)$. Consider *PPI* q_j which is a predecessor of q_i. By definition $POW(q_j)$ gives the lower bound on the power cost of all *PPIs* containing q_j. Therefore the minimum power cost for all predecessor *PPIs* of q_i gives the lower bound on the power cost of all *PPIs* containing q_i. If $POW(q_i)$ is less than the power cost of all its predecessor *PPIs* then its power cost must be less than the power cost of all *PPIs* containing it which means q_i is a *PPI* itself and this will contradict our initial assumption.

■

Theorems 3.1 and 3.2 suggest a recursive procedure for generating the set of all *PPIs* for a function. In this procedure, efficiency is achieved by using the set of *PPIs* with *n* literals for generating and checking candidate *PPIs* with *n+1* literals. The input to this procedure is the set of all prime implicants of the function. The procedure will then generate set *PP* of sets PP_n where each PP_n will contain *PPIs* with *n* literals.

```
1:      function Generate_PPI(F)
2:  F is a Boolean function with input set V = (v_1,...v_N)
3:  begin
4:      P = generateAllPrimes (F)
5:      PP = initializePP(P)
6:      for ( i = 1 ; i < N ; i++) do
7:          foreach ( implicant q with at least one predecessor in PP_i ) do
8:              q_j = findMinPowPredecessor(PP_i)
9:              l = literalLoweredInQ(q, q_j)
10:             if ( sp(l) < (sp(q_j) - sp^2(q_j) )/(1+ sp^2(q_j) ) ) then
11:                 PP_{i+1} = PP_{i+1} ∪ q
12:             endif
13:         endfor
14:     endfor
15:     return PP
16: end
```

Figure 3.2 Computing the set of all *PPIs*

Procedure *initializePP* is used to initialize *PP* by placing prime implicants with n literals in PP_n. The main loop in the procedure updates PP_{i+1} by adding all implicants that have a predecessor cube in PP_i and also satisfy the conditions for being a *PPI*. In this loop *findMinPowPredecessor* returns the predecessor cube of q which has the smallest power cost. Procedure *literalLoweredInQ* will return the literal which was lowered in q_j to obtain q. Set *PP* returned by the procedure contains the set of all *PPIs* of the function.

3.5 Exact Minimization Algorithm for Low Power

Once the set of all *PPIs* of the function are generated, a minimal covering problem will be used to select a set of *PPIs* which cover the function and have minimum power cost. The complexity of the minimum covering problem grows exponentially as a function of the number of implicants which are being considered. This means that even though it might be possible to find an exact minimum area solution for a function, finding an exact minimum power solution might not be possible due to the increased number of implicants being considered. The following section presents an analysis of the number of *PPIs* that will be generated as a function of the number of prime implicants of the function. This analysis will show that, in general, the total number of *PPIs* will not be significantly larger than the number of prime implicants of the function. The experimental results presented at the end of the chapter will confirm the results of this analysis.

3.6 Upper Bounds on the Expected Number of PPIs

Two issues need to be considered while including *PPIs* in the optimization procedure for a function f. The first issue is the process of generating the set of all *PPIs* for the function. This issue was analyzed in detail in the previous section and a procedure for finding all *PPIs* was proposed. The second and more important issue is the increased complexity of the problem due to including *PPIs* in the set of candidate implicants in the covering problem. Solving the covering problem in two-level function optimization will become more difficult as the number of candidate implicants are increased. A function f with N inputs has 3^N possible implicants. This means that including *PPIs* in the covering problem may increase the number of candidate implicants therefore resulting in an unacceptable increase in the run-time of the covering problem. This section presents an analysis of the expected number of *PPIs* generated while optimizing a function for power. Based on this analysis, methods are proposed to selectively control the expected number of *PPIs* which are generated in the process.

3.6.1 Implicant signal probabilities as random variables

The *PPIs* generated for a boolean function depends on the signal value statistics of the inputs to the function. This means that an effective analysis of the expected number of generated *PPIs* should consider the input activity behavior. In the following, basic definitions for the statistical properties of signal values are provided. These definitions are then used to derive a formula for the expected number of generated *PPIs*.

Definition 3.3 *The distribution function G of random variable Y is the function G(y) = P(A_y) where A_y is the event (R is the set of real numbers):*

$$A_y = \{w : Y(w) \le y\}, y \in R$$

Definition 3.4 *The density function g of random variable X is the function g(x) such that:*

$$G(y) = \int_{-\infty}^{y} g(v)dv \tag{3.7}$$

Definition 3.5 *The sample space for the signal probability of a set of inputs is the set of points (Q:[0,1]). The random variable X(Q) is defined as the probability of a randomly selected input having a signal probability $q \in Q$. Assuming that given a set of inputs, all signal probabilities are equally likely, the distribution and density functions for X are given as follows:*

$$G(y) = \begin{cases} 0 & y < 0 \\ y & 0 \le y \le 1 \\ 1 & y > 1 \end{cases} \tag{3.8}$$

$$g(y) = \begin{cases} 1 & 0 < y < 1 \\ 0 & otherwise \end{cases} \tag{3.9}$$

In what follows, spatial and temporal independence is assumed at the inputs to the function.

Lemma 3.3 *Given an implicant q with (n+1) literals and p(q) = x, then:*

$$g(x) = \frac{(\ln(1/x))^n}{n!} \tag{3.10}$$

Proof: Given $z = a.b$, $g(z)$ is computed as follows [11]:

$$g(z) = \int_{-\infty}^{\infty} \frac{1}{|u|} g_{a,b}(u, z/u) du \tag{3.11}$$

Variables a and b are assumed to be independent. Using equation 3.9:

$$g(z) = \int_{z}^{1} \frac{1}{u} g_a(u) g_b(z/u) du = \int_{z}^{1} \frac{1}{u} du = \ln(1/z)$$

Now given $(w = z.c)$, equations 3.10 and 3.11 are used to find $g(w)$:

$$g(w) = \frac{(\ln(1/z))^2}{2}$$

Same approach is then used to derive the density function for implicants with more literals. The claim of the proof is then proved by deduction.

■

Lemma 3.4 *Given an implicant q with (n+1) literals and p(q) = x, then:*

$$G(x) = x \cdot \sum_{k=0}^{n} \frac{(\ln(1/x))^k}{k!} \tag{3.12}$$

Proof: Using equation 3.7:

$$G(x) = \int_{-\infty}^{x} g(u) du = \int_{0}^{x} \frac{(\ln(1/u))^n}{n!} du$$

Solving this equation using variable substitution and the following identity proves the lemma.

$$\int u^n e^{\lambda u} du = e^{\lambda u} \sum_{m=0}^{n} \frac{(-1)^m}{\lambda^{m+1}} \frac{n!}{(n-m)!} u^{(n-m)}$$

■

Using the equations developed for the distribution and density of a product term, we now present an analysis which will provide us with an upper bound on the expected number of *PPIs* that will be generated assuming a uniform density for the input signal probabilities.

3.6.2 Upper bounds on the number of generated *PPIs*

As was stated previously, lowering literal l in a *PPI* q_i with $sp(q_i)=x$ will result in a *PPI* if and only of the condition in equation 3.4 is satisfied. Therefore, the probability of creating a new *PPI* q_j is the probability of condition in equation 3.4 being true. Given an implicant q with $sp(q)=x$ and a literal l to be raised in q, equation 3.13 shows that $g(sp_l<x)$ gives an upper bound on the probability of the condition in equation 3.4 being true. $g(sp_l<x)$ can then be computed by finding $g(x-sp_l>0)$.

$$\left(\frac{x-x^2}{1+x^2} < x(1-x) < x\right) \Rightarrow g\left(sp_l < \frac{x-x^2}{1+x^2}\right) < g(sp_l < x(1-x)) < g(sp_l < x) \quad (3.13)$$

The following lemma provides a method for computing $g(sp_l<x)$.

Lemma 3.5*Given the input signal probabilities are uniformly distributed and un-correlated, an upper bound on the probability of generating a PPI with n+1 literals is given by* $(0.5)^n$ *for* $n>0$.

Proof: We first find the density function for the difference of two random variables. We then use this density function to find the probability of the difference being positive.

Given two random variables X and Y, the density function for random variable Z=X-Y is [11]:

$$g(z) = \int_{-\infty}^{\infty} g_{x,y}(u, u-z) du \quad (3.14)$$

Our goal is to find $g(z=x-sp_l)$ therefore $g(x)$ is given by equation 3.10 and g(y) by equation 3.9. Using these equations and the independence between X and Y:

$$g(z) = \begin{cases} \int_0^{(1+z)} g_x(u)du & -1 \leq z < 0 \\ \int_z^1 g_x(u)du & 0 \leq z < 1 \\ 0 & otherwise \end{cases}$$

Then $g(z)$ for positive values of z where x is the output signal probability of an implicant with $(n+1)$ literals is computed as:

$$g(z) = \int_z^1 \frac{(\ln(1/u))^n}{n!} du = 1 - z \cdot \sum_{k=0}^{n} \frac{(\ln(1/z))^k}{k!} \quad (3.15)$$

The probability of $z>0$ is then computed by integrating equation 3.15 for all positive values of z:

$$p(z > 0) = \int_0^1 \left(1 - z \cdot \sum_{k=0}^{n} \frac{(\ln(1/z))^k}{k!}\right) dz = 0.5^{n+1}$$

Equation 3.14 gives the upper bound for the probability of generating a *PPI* with $n+2$ literals for $n \geq 0$. Therefore the upper bound for the probability of generating a *PPI* with $(n+1)$ literal is given by 0.5^n for $n > 0$.

■

The result of lemma 3.5 provides an upper bound on the probability of generating a *PPI*. The following lemma provides a tighter bound for the probability of generating a *PPI*.

Lemma 3.6*Given that the input signal probabilities are uniformly distributed and un-correlated, the following equation presents an upper bound on the probability of generating a PPI with (n+1) literals:*

$$\left(\frac{1}{2}\right)^n - \left(\frac{1}{3}\right)^n \quad (3.16)$$

Proof: Assume an implicant q with n inputs and with output signal probability x and literal l to be lowered in q to generate a *PPI*. Here we find $g(sp_l<x(1-x))$ which provides a better upper bound for the probability of generating a *PPI*. Assuming random variable $T = X(1-X)$, $g(t)$ is computed using equations 3.7 and 3.12:

$$g(t) = \frac{1}{n!} \cdot \frac{1}{\sqrt{1-4t}} \cdot \left[\left(\ln\frac{2}{1+\sqrt{1-4t}}\right)^n + \left(\ln\frac{2}{1-\sqrt{1-4t}}\right)^n\right] \tag{3.17}$$

Our goal is to find $g(z=t-sp_l)$ therefore $g(x)$ is given by equation 3.17 and $g(y)$ by equation 3.9. Using these equations and the independence between T and Y, $g(z)$ for positive values of z is computed as follows:

$$g(z) = \begin{cases} \int_z^{1/4} g_t(u)du & 0 \le z < 1/4 \\ 0 & 1/4 \le z \end{cases}$$

Then, using equation 3.17, $g(z)$ for positive values of z where $t=x(1-x)$ and x is the output signal probability of an implicant with $(n+1)$ literals for $1/4<z$ is computed as:

$$\left(\frac{1+\sqrt{1-4t}}{2}\right) \cdot \sum_{k=0}^{n} \left(\frac{\left(\ln\frac{2}{1+\sqrt{1-4t}}\right)^k}{k!}\right) - \left(\frac{1-\sqrt{1-4t}}{2}\right) \cdot \sum_{k=0}^{n} \left(\frac{\left(\ln\frac{2}{1-\sqrt{1-4t}}\right)^k}{k!}\right) \tag{3.18}$$

The probability of $z>0$ is then computed by integrating equation 3.18 for all positive values of z:

$$p(z>0) = \int_0^{1/4} g(z)dz = \left(\frac{1}{2}\right)^{n+1} - \left(\frac{1}{3}\right)^{n+1} \tag{3.19}$$

Equation 3.19 gives the upper bound for the probability of generating a *PPI* with $n+2$ literals for $n \ge 0$. Therefore the upper bound for the probability of generating a *PPI* with $(n+1)$ literal is given by equation 3.16.

■

The following lemma gives the expected number of *PPI*s included in the covering problem due to a prime implicant with n literals.

Lemma 3.7*Given a prime implicant q <u>with n literals</u> for a function f (with N inputs), an upper bound on the expected number of PPIs introduced due to this prime implicant is given by:*

$$(N-n)0.5^{n-1}(2^{N-n}-1) \qquad n \geq 0 \tag{3.20}$$

Proof: The number of implicants with $n+m$ literals which are contained in q is given by:

$$2^m \binom{N-n}{m}$$

Using lemma 3.5, the expected number of *PPIs* with the same number of literals, generated due to q is:

$$\sum_{j=1}^{N-n} 2^j \cdot \binom{N-n}{j} \cdot 0.5^{n+j-1} = 0.5^{n-1} \cdot (2^{N-n}-1) \tag{3.21}$$

Multiplying this equation with *(N-n)* the number of different size implicant contained in q proves the claim of this lemma.

■

Lemma 3.8*The upper bound on the E(PPI), expected number of PPIs introduced due to each prime implicant q (with any number of literals) of a function f with N inputs, is given by:*

$$N \cdot \frac{2}{3} \cdot \left(2 - \left(\frac{2}{3}\right)^{N-1}\right) \tag{3.22}$$

Proof: Assuming all Boolean functions are equally probable, the probability of having a prime implicant with n literals for a function with N inputs is given by:

$$\frac{2^n \binom{N}{n}}{3^N} \tag{3.23}$$

Then using equation 3.20,

$$E(PPI) = \sum_{n=0}^{N} \frac{2^n \binom{N}{n}}{3^N} (N-n)0.5^{n-1}(2^{N-n}-1) \tag{3.24}$$

Solving this equation will lead to the claim of this lemma.

■

Note that the value in equation 3.24 is a loose bound. This is because lemma 3.5 gives an upper bound on the expected probability of generating a *PPI*. The bound in equation 3.24 can be improved by using the better bound provided by lemma 3.6. In addition, the bound given in the previous lemma does not consider the fact that some *PPIs* will be contained in more than one prime implicant. This means that such *PPIs* will be accounted for more than once.

Equation 3.22 shows that for large N the number of *PPIs* introduced for each prime can potentially increase as a linear function of the number of inputs, therefore exponentially increasing the complexity of the covering problem. The following lemma gives an upper bound on the expected number of *PPIs* for each prime implicant if we only choose to raise m literals in each prime implicant.

Lemma 3.9 *If only less than or equal to m literals are lowered for each prime implicant, then the upper bound on the expected number of PPIs introduced for each prime implicant for a function with N inputs is given by:*

$$\frac{m(2^m - 1)}{2}\left(\frac{2}{3}\right)^N \tag{3.25}$$

Proof: This lemma is proved by changing the limits of the summation in equation 3.21 so that variable j changes from *1* to m. Note again that the value given in this lemma is an upper bound since in addition to previous approximations, it assumes that the number of literals in a *PPI* can increase beyond the number of function inputs.

■

This lemma can be used to control the expected number of *PPIs* generated for each prime implicant. For example having knowledge of N, we can select a value for m such that equation 3.25 evaluates to a number less than 1.

The drawback of this method, however, is that it cannot easily be incorporated into the procedure presented in the previous section for generating the set of all *PPIs*. At the same time, it is desirable to raise literals in implicants with a lower number of literals since this will, in general, result in more reduction in power consumption at the output of the gate implementing this implicant. An alternative approach to increasing the literals of each prime implicant by m literal at most as suggested by lemma 3.9, is to set an absolute maximum number of literals for all *PPIs* in order to control the expected number of *PPIs*. This method will give more priority to implicants with lower number of literals.

Lemma 3.10 *Given a Boolean function f with N inputs, if only PPIs with up to m literals are considered, then the expected number of PPIs introduced for each prime implicant is given by:*

$$\frac{2}{3^N} \cdot (m \cdot 2^m \cdot 1.5^N - m \cdot 2^m - m \cdot 3^{m-1} + N \cdot 2^{N-1}) \tag{3.26}$$

Proof: The proof is obtained by changing the limits of the summation in equation 3.24 so that it changes from 0 to m. The claim of this lemma is then proved by using the following relations to find an upper bound to the summation where the index variable changes from *0* to m.

$$\left(\left(\sum_{j=0}^{N} 2^{-j} \cdot \binom{N}{j}\right) \geq \left(\sum_{j=0}^{m} 2^{-j} \cdot \binom{N}{j}\right)\right) \quad , \quad \left(\left(\sum_{j=0}^{m} 2^{-j} \cdot \binom{N}{j}\right) \geq \left(\sum_{j=0}^{m} 2^{-j} \cdot \binom{m}{j}\right)\right)$$

■

Using the equation given in lemma 3.10, we can selectively choose m to control the expected number of *PPIs* generated for each prime implicant. Table 3.1 shows values for m that will result in the expected increase in the number of *PPIs* to be 1 and 10. For example, given a function with 30 inputs, the expected number of *PPIs* generated for each prime implicant will be less than 1 if we only consider *PPIs* with 24 or less literals. This lemma can be used to directly control the procedure for computing *PPIs* by only processing PP_k where $k<m$.

This section used the concepts of density and distribution functions to find upper bounds on the expected number *PPIs* which are introduced in the optimization problem. Using this upper bound, a method was described for limiting the number of generated *PPIs*. The next section discusses function minimization for power and issues that affect the quality of the minimum power solution.

N	*E(PPI)=1*	*E(PPI)=10*
5	1	5
10	6	9
15	10	13
20	15	18
25	19	23
30	24	27
35	29	32
40	33	37

Table 3.1 Maximum literals to be raised as a function of number of inputs

3.7 Exact Function Minimization for Low Power

Solving the exact two-level function minimization problem for area consists of first generating a set of prime implicants that will be considered as candidate cubes of the function, and then selecting a subset of these primes which results in minimum cost of the cover. The set of primes of a function can be partitioned into three classes. *Essential primes*, *partially redundant primes* and *totally redundant primes* [1]. Essential primes are defined as primes that cover at least one minterm that is covered by only one prime implicant. Totally redundant primes are then defined as the set of primes which are covered by the essential primes of the function. The remaining primes of the function are then defined as partially redundant primes. In order to study the complexity of the two-level minimization procedure, a function can be classified as *trivial*, *non-cyclic* or *cyclic* [9]. For a trivial function, all primes of the function are essential. Therefore, the two-level representation of the function includes all prime implicants of the function. A non-cyclic function does not have any partially redundant primes. Consequently, the two level representation of the function again consists only of the essential primes of the function since all totally redundant primes can be dropped by including the set of all essential primes. The two-level representation of a cyclic function includes all essential primes, and a subset of the partially redundant primes of the function. Note that for trivial and non-cyclic functions the minimum solution is obtained by first generating the set of all primes and then partitioning the primes into essential, partially redundant, and totally redundant. It is not necessary to solve a minimum covering problem for trivial and non-cyclic functions. Cyclic functions do, however, require that a minimum covering problem be solved to minimize the cost function under consideration.

The classification of prime implicants and functions based on the makeup of their prime implicants can also be applied to functions when power is being minimized. *Essential PPIs*, *Totally Redundant PPIs*, and *Partially Redundant PPIs* are defined similar to their counterparts, while optimizing area. Essential *PPIs* are *PPIs* that cover a minterm that is covered by only one *PPI*. Totally redundant *PPIs* are *PPIs* that are covered by essential *PPIs*. The remaining *PPIs* are defined as partially redundant *PPIs*. *Power trivial* functions are defined as functions where all *PPIs* are essential. *Power non-cyclic* functions are defined as functions that have no partially redundant *PPIs*, and *power cyclic* functions are functions which have partially redundant *PPIs*.

Figure 3.3 shows the relationship between function classifications for area and power. As shown in the figure, a trivial function can be power trivial, power non-cyclic, or power cyclic. However, a cyclic function can only be a power cyclic function. This relationship between function classification is due to non-prime impli-

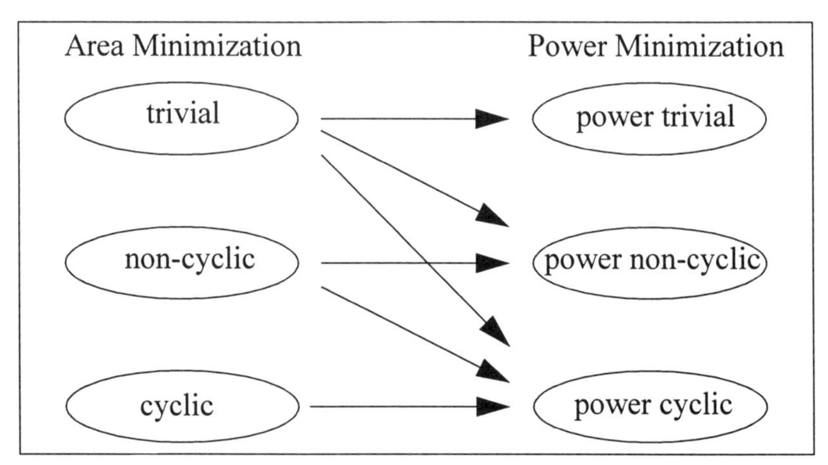

Figure 3.3 function characterization for low power and low area optimizations

cants which are included in the set of *PPIs* of the function. Note that if a trivial function is also power trivial then in order to guarantee functional correctness, any power optimization and area optimization algorithm will produce the same results. In general, it is possible to obtain a lower power solution for a trivial or non-cyclic function if this function is a power cyclic function.

Example 3.2: (example 3.1 revisited)

Function F as defined in example 3.1 is a trivial function with two essential primes *(a.b)* and *(b.c)*. This means that the minimum area solution for this function is readily obtained from the set of essential primes. This function, however is a power cyclic function since it has one essential *PPI*: *(a.b)* and two partially redundant *PPIs*: *(b.c)* and $(\bar{a}.b.c)$.

An important observation is that power trivial and power non-cyclic functions all have the same minimum solutions for area and power. This means that for such functions a minimum area solution will also provide the minimum power solution. Therefore, we only need to optimize power cyclic functions for power.

There are two possible sources of power savings while optimizing a power cyclic function. First, if the power cyclic function is also a cyclic function, then the freedom to choose different subsets of the partially redundant primes of the function can potentially lead to power reductions. This means that it is possible to obtain power savings in a cyclic function even if no non-prime *PPIs* are included in the covering problem. This is illustrated in the following example:

Example 3.3:

Consider the cyclic function F with essential primes $(b.\bar{c})$ and $(\bar{b}.c)$ and partially redundant primes $(a.b)$ and $(a.c)$ and $sp(a)=0.2$, $sp(b)=0.9$ and $sp(c)=0.8$. The two possible solutions for F are:

$$F_1 = b.\bar{c} + \bar{b}.c + a.b \qquad F_2 = b.\bar{c} + \bar{b}.c + a.c$$

Then:

$$Pow(F_1) = 2.35 \qquad Pow(F_2) = 2.72$$

This example shows that it is possible to obtain reductions in power even if only prime implicants of the function are included in the minimum covering problem. In this example a minimum area procedure would not differentiate between the two possible implementations of the function. A power optimization procedure, however, will choose the lower power solution.

As shown in example 3.3, the second source of power savings is due to inclusion of *PPIs* which are not prime implicants of the function.

3.8 Experimental Results

The methods described in this chapter have been implemented as an extension of ESPRESSO-EXACT. Options have been built in to allow for minimizing a function for area, minimizing a function for power using only the set of prime implicants of the function, and finally, minimizing the function for power using the set of all power prime implicants of the function. The following experiments are used to measure the effectiveness of the optimization process.

In addition, a procedure has also been implemented where given the number of inputs and the ratio of the number of on-set to number of all points in the space of the function (i.e. on-set ratio), will create a function where the on-set points are randomly set to achieve the required on-set ratio. Using this procedure, we created 9 random functions of n inputs with the on-set ratios of R with n ranging from 3 to 9 and R having values in the set {0.2, 0.4, 0.7, 0.95}. For each random function of n inputs and on-set ratio R, 7 different sets of uniformly distributed input signal probabilities were obtained by randomly setting the input signal probabilities. The set of 9 random functions with 7 different sets of input signal probabilities provided 63 different functions of n inputs with on-set ratio of R. The exact simplification procedure developed in this chapter was then applied to all these functions. The results of these experiments are presented in the following tables.

Table 3.3 shows the relationship between the number of prime implicants and power prime implicants of the functions. Columns 1 and 2 give the number of inputs and the on-set ratio of the functions. Columns 3, 4, 5 show total number of essential, partially redundant, and totally redundant primes for the 63 functions represented by each row. Column 6 shows the total number of prime implicants for all 63 functions. Columns 7, 8, and 9 show the number of essential, partially redundant and totally redundant power prime implicants of the function. Column 10 shows the total number of *PPIs* for all 63 functions. Column 11 gives the ratio of the total number of *PPIs* to the total number of prime implicants for the 63 functions. In our implementation of the *PPI* generation algorithm, we always generate all *PPIs* for the function. The results also show that the increase in number of *PPIs* of the function is, in general, larger for functions with higher on-set ratio. This relationship can be observed by noting that the ratio given in column 11 increases as the on-set ratio of the functions with the same number of inputs increases.

Table 3.2 shows the results of the power optimization on the example circuits. Column 3 gives the sum of the power for all 63 functions represented by that row when the functions are optimized for area. Column 4 gives the total power when the functions are optimized for power using only the prime implicants of the function. Column 6 gives the total power for all 63 functions when the functions are optimized for power using the set of all *PPIs* of the function. The exact covering problem for functions with 8 and 9 inputs could not be solved. Therefore, the results in this table for functions with 8 and 9 inputs are obtained by solving the minimum covering problem heuristically. The results in this table show that for function with large on-set ratio, it is possible to achieve more than 10% reduction in power. This is due to the fact that the prime implicants of functions with larger on-set ratio are, in general, larger (with fewer literals) and as predicted by lemma 3.6, more *PPIs* will be generated for prime implicants with smaller number of literals. Note that for functions with 8 and 9 inputs, there was no reduction in power even for functions with high on-set probability. Note that for these functions, the solution provided here does not represent the minimum solution since a heuristic approach was used in solving the minimum covering problem in these cases.

ex		area	power (no *PPIs*)		power (with *PPIs*)	
inputs	on-set ratio	power	power	gain	power	gain
1	2	3	4	5	6	7
3	0.20	53.77	53.77	1.00	53.77	1.00
3	0.40	125.35	125.35	1.00	124.73	1.00
3	0.70	135.45	132.89	0.98	132.15	0.98
3	0.95	56.00	56.00	1.00	56.00	1.00
4	0.20	204.30	204.30	1.00	204.30	1.00
4	0.40	287.01	285.84	1.00	283.06	0.99
4	0.70	261.08	256.22	0.98	255.35	0.98
4	0.95	72.58	72.58	1.00	72.58	1.00
5	0.20	326.20	326.20	1.00	324.94	1.00
5	0.40	395.25	392.21	0.99	390.48	0.99
5	0.70	498.12	492.11	0.99	492.11	0.99
5	0.95	136.22	134.79	0.99	128.83	0.95
6	0.20	779.57	778.58	1.00	777.71	1.00
6	0.40	970.40	952.26	0.98	950.23	0.98
6	0.70	925.78	874.86	0.94	872.99	0.94
6	0.95	329.75	307.73	0.93	295.42	0.90
7	0.20	1587.57	1576.39	0.99	1576.37	0.99
7	0.40	1950.88	1915.36	0.98	1913.77	0.98
7	0.70	1964.64	1886.00	0.96	1881.04	0.96
7	0.95	570.27	524.29	0.92	507.97	0.89
8	0.20	3843.66	3821.99	0.99	3822.19	0.99
8	0.40	4270.36	4243.75	0.99	4242.33	0.99
8	0.70	3373.67	3331.54	0.99	3334.61	0.99
8	0.95	1151.54	1118.35	0.97	1110.86	0.96
9	0.20	6740.49	6715.59	1.00	6714.56	1.00
9	0.40	8921.95	8927.40	1.00	8932.19	1.00
9	0.70	8456.69	8576.90	1.01	8587.87	1.02
9	0.95	1927.16	1852.17	0.96	1834.45	0.95
Heuristic solution for the covering problem			exact solution of the minimum covering problem.			

Table 3.2 Power optimization for two-level functions

ex		Prime implicants				Power Prime Implicants				
inputs	on-set ratio	Essent.	Part. Redund.	Tot. Redund.	SUM	Essent.	Part. Redund.	Tot. Redund.	SUM	ratio 10/6
1	2	3	4	5	6	7	8	9	10	11
3	0.20	63	0	0	63	63	0	0	63	1.00
3	0.40	126	0	0	126	121	10	23	154	1.22
3	0.70	126	56	0	182	110	88	7	205	1.13
3	0.95	189	0	0	189	189	0	0	189	1.00
4	0.20	154	0	0	154	154	0	7	161	1.05
4	0.40	252	28	0	280	223	87	12	322	1.15
4	0.70	168	231	35	434	139	337	34	510	1.18
4	0.95	252	0	0	252	252	0	0	252	1.00
5	0.20	280	0	14	294	273	14	16	303	1.03
5	0.40	322	203	35	560	297	275	49	621	1.11
5	0.70	357	294	77	728	289	525	94	908	1.25
5	0.95	252	126	0	378	223	246	91	560	1.48
6	0.20	525	49	28	602	515	70	36	621	1.03
6	0.40	490	707	35	1232	462	795	60	1317	1.07
6	0.70	217	1981	21	2219	179	2237	72	2488	1.12
6	0.95	140	959	7	1106	90	1676	91	1857	1.68
7	0.20	889	350	70	1309	884	362	72	1318	1.01
7	0.40	875	1561	161	2597	843	1694	177	2714	1.05
7	0.70	217	4235	0	4452	192	4778	39	5009	1.13
7	0.95	70	2674	0	2744	32	4615	71	4718	1.72
8	0.20	1533	1134	140	2807	1523	1154	148	2825	1.01
8	0.40	1323	4116	182	5621	1295	4249	201	5745	1.02
8	0.70	245	10514	7	10766	211	11633	26	11870	1.10
8	0.95	56	6706	0	6762	22	10000	43	10065	1.49
9	0.20	3080	1897	413	5390	3052	1966	412	5430	1.01
9	0.40	1526	10696	161	12383	1488	10953	177	12618	1.02
9	0.70	203	24885	7	25095	187	25856	20	26063	1.04
9	0.95	42	18186	0	18228	30	24746	46	24822	1.36

Table 3.3 Number of prime implicants and power prime implicants

References

[1] R. Brayton, G.D. Hachtel, C. McMullen and A. Sangiovanni-Vincentelli. "Logic Minimization Algorithms for VLSI Synthesis." *Kluwer Academic Publishers*, Boston, 1984.

[2] S. Devadas, K. Keutzer and J. White. "Estimation of Power Dissipation in CMOS Combinational Circuits using Boolean Manipulations." In *IEEE Transactions on Computer Aided Design*, vol 11, no 3, March 1992.

[3] H. Fleisher, and L. Maissel. "An introduction to array logic," IBM Journal of Research and Development 19, March 1975, 98-109.

[4] Z. Kohavi. "Switching and Finite Automata Theory, second ed." McGraw-Hill, New Yoro, 1978.

[5] R. Marculescu, D. Marculescu, M. Pedram. "Logic level power estimation considering spatio-temporal correlations." In proceedings of the *IEEE International Conference on Computer Aided Design*, pages 294-299, 1994.

[6] E. J. McCluscky. "Minimization of Boolean functions," Bell Syst. Technical Journal (Nov. 1956), 1417-1444.

[7] W. Quine. "The problem of simplifying truth functions," American Math. Monthly 59, 1952, 521-531.

[8] W. Quine. "A way to simplify truth functions," American Math. Monthly 62 (Nov. 1995),627-631.

[9] R. Rudell. "Logic Synthesis for VLSI Design." Ph.D. thesis, University of California, Berkeley, 1989.

[10] A. A. Shen, A. Ghosh, S. Devadas, and K. Keutzer. "On average power dissipation and random pattern testability of CMOS combinational logic networks." In proceedings of the *IEEE International Conference on Computer Aided Design*, November 1992.

[11] D. Stirzaker. "Elementary Probability." Cambridge University Press, 1994.

[12] S. B. K. Vrudhula, H.Y. Xie. "Techniques for CMOS Power Estimation and Logic Synthesis for Low Power." In proceedings of the *International Workshop on Low Power Design*, pages 21-26, 1994.

CHAPTER 4 Two-Level Logic Minimization in PLAs

Random logic and *structured logic* provide two distinct approaches for implementing digital logic circuits. Random logic implementations, in general, provide lower silicon area utilization and also faster operation times. On the other hand, structured logic implementations allow for a shorter design cycle and easier testing of the implemented logic. Depending on the design requirements, a mix of structured and random logic is used to implement today's digital circuits.

Programmable Logic Arrays (PLAs) provide a regular and structured way to implement two-level Boolean functions. Normally a PLA can realize several output functions concurrently. The PLA structure can be realized in either CMOS or NMOS technology. In either case, a PLA consists of two major subsections or planes (see figure 4.1). One is the AND plane, which requires double-rail inputs to generate the product terms required by the defining logic functions. The other is the OR plane which combines the product terms to generate the output functions. A convenient measure of a PLA's size is the triplet (i, p, o) where i is the number of inputs, p is the number of product terms and o is the number of outputs. The number of potential transistors in the AND and OR planes is given by the expression $(2i + o)p$. Increasing i or o adds to the width of the AND plane or the OR plane, respectively. Increasing p adds to the height of both the AND and OR planes. A relative measure of the PLA size is then given by the calculation $(2i+o)p$. Since for a given set of Boolean functions, the number of inputs and outputs is constant, the problem of optimizing a PLA for area is equivalent to minimizing the number of product terms necessary to implement the Boolean functions.

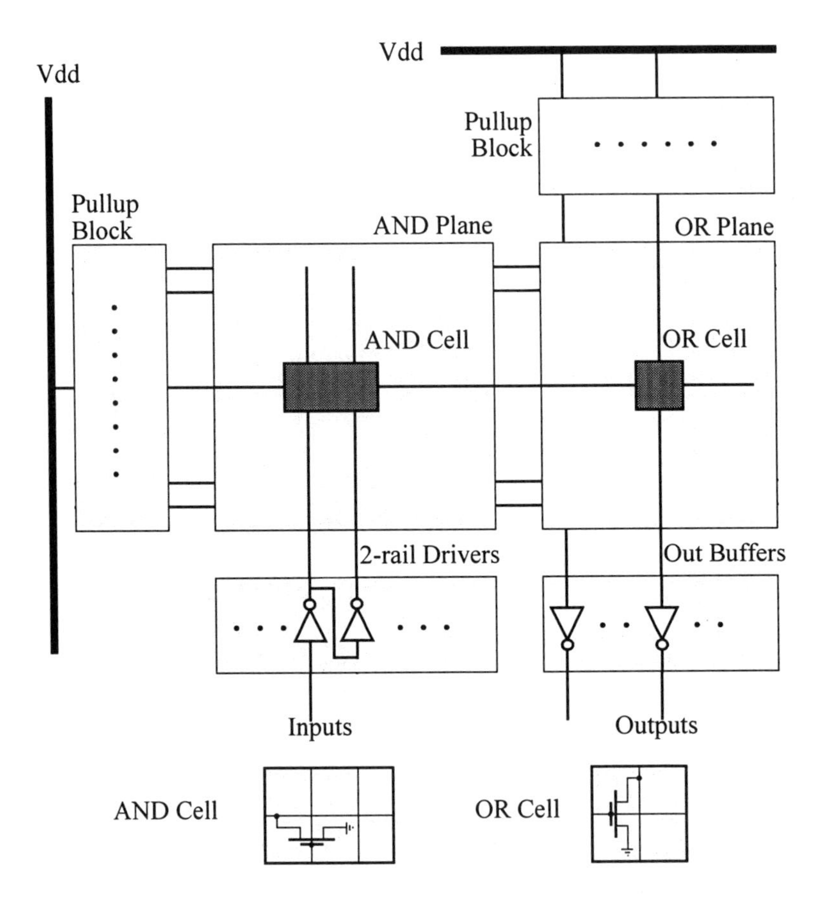

Figure 4.1 PLA Architecture

This chapter addresses the problem of minimizing power consumption in PLA implementations of two-level Boolean functions. The loading information in PLAs are known for a specific technology/implementation and hence the switched capacitance can be minimized directly. In particular, this chapter shows how logic minimization techniques for area are modified to obtain a minimum power solution.

This chapter is organized as follows: Section 4.1 discusses the power model used in minimizing the power consumption in PLAs. Section 4.2 shows how prime implicants correspond to implicants needed for power optimization for both pseudo-NMOS and dynamic PLA implementations. Section 4.3 will then use the results in section 4.2 to minimize power consumption in PLAs. Experimental results are presented in section 4.4.

4.1 Power Model for PLA Logic Implementation

Normally, high speed PLAs are built by transforming the SOP representation of a two level logic to the NOR-NOR structure with inverting inputs and outputs and implementing it with two NOR arrays (Figure 4.2). Two common types of implementing NOR arrays are pseudo-NMOS NOR gates and dynamic CMOS NOR gate. It is assumed here that functions are implemented using a PLA that is driven using static CMOS drivers. This means that the cost function for an implicant in the final cover of the function assumes that the AND gate is implemented using either pseudo-NMOS or dynamic technology while the input power consumption follows the power cost model for static CMOS structures. Each input literal of AND term in the final implementation of the function creates additional load on the input CMOS drivers. Equation 1.1 will be used to measure the power contribution of the PLA input literals on the total power consumption.

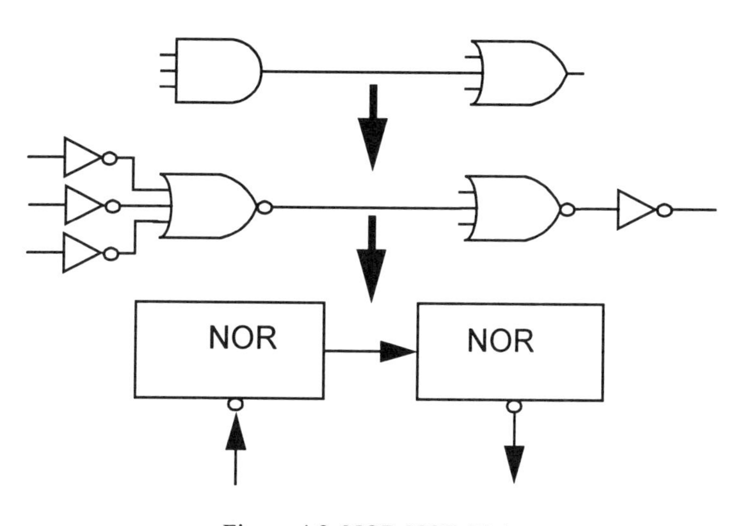

Figure 4.2 NOR-NOR PLA

In a pseudo-NMOS NOR gate, static power dissipation is the primary source of power consumption (see figure 4.3). When a pseudo-NMOS NOR gate evaluates to zero, both the PMOS and NMOS parts of the gate are conducting and there exists a direct current path. The charging and discharging energy is small compared with that dissipated by the direct current when the frequency of operation is not extremely large. Furthermore, the direct current I_{dc} is relatively constant regardless of the number of NMOS transistors that are on. It is assumed here that static power dominates in pseudo-NMOS PLAs. The power cost at the output of a product term p is given by:

$$OutPower(p) = V_{dd} \cdot I_{dc} \cdot sp_p^0 \tag{4.1}$$

where sp^0_p is the probability that the product term p evaluates to 0. The total power cost of a product term which accounts for power consumptions at its inputs and output is then given by:

$$Power(p) = \frac{V_{dd}^2 \cdot f}{2} \cdot \sum_{i=1}^{k} c_i sw_i + V_{dd} \cdot I_{dc} \cdot sp_p^0 \tag{4.2}$$

where c_i gives the load due to product term p on input i.

In dynamic PLA circuits (see figure 4.5), dynamic power consumption is the major source of power dissipation. The output of a product term is pre-charged to 1

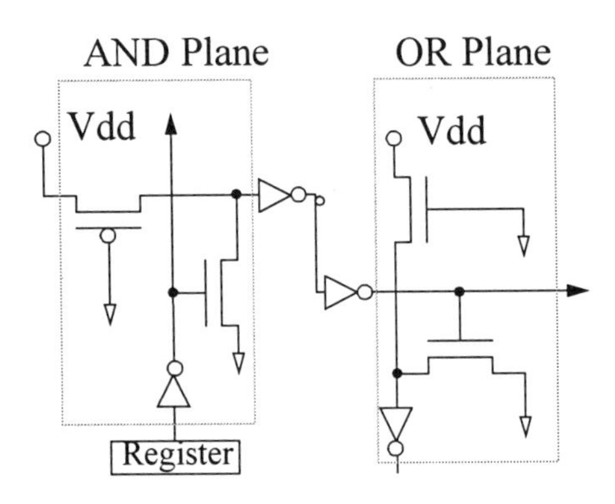

Figure 4.4 Pseudo-NMOS PLA

and switches when it evaluates to 0. Therefore, the power consumption at the output of a product term p is given by:

$$OutPower(p) = \frac{V_{dd}^2 \cdot f}{2} \cdot c_p \cdot sp_p^0 \tag{4.3}$$

where c_p is the load seen at the output of the product term and is estimated as the number of gates in the OR-plane that p fans out to.

The total power cost due to a product term p is obtained by taking into account the power consumption on the inputs and clocks and is given by:

$$Power(p) = \frac{V_{dd}^2 \cdot f}{2} \cdot \left(\sum_{i=1}^{k} c_i sw_i + C_p sp_p^0 + c_{clock} \times 2 \right) \tag{4.4}$$

where c_{clock} is the load capacitance of the pre-charge and evaluate transistors that the clock drives.

The power due to the OR plane of the PLA is measured using equations 4.1 and 4.3 presented for the product terms since both the input and output planes of the PLA are implemented using NOR gates.

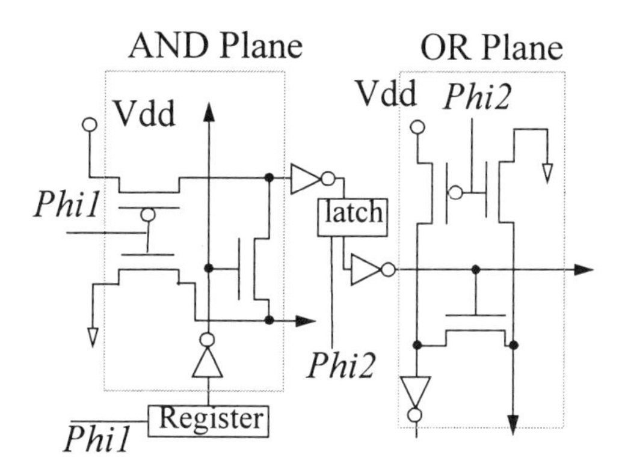

Figure 4.5 Dynamic PLA

4.2 Prime implicants and PLA Power Optimization

The two level logic minimization problem for low power is equivalent to finding a cover *C(F)* such that the following objective function is minimized:

$$\sum_{p \in C(F)} Power(p) + \sum_{q \in O(F)} OutPower(q) \qquad (4.5)$$

where *O(F)* is the set of gates in the OR plane.

As stated before, the set of prime implicants of the function is sufficient for finding a minimum area solution. The following two sections examine whether this statement is also true for power minimization. Cases for pseudo-NMOS and dynamic PLAs are discussed separately.

4.2.1 Pseudo-NMOS PLA

Don't care conditions provide flexibility in optimizing Boolean functions. It is, therefore, important to consider the effect of using don't care conditions during the optimization process. The first level NOR gate in a pseudo-NMOS PLA will draw direct current when the AND term evaluates to 0. Similarly, the second level NOR gate will draw direct current when the gate output is 0 (i.e., the function output evaluates to 1). Including don't care in the cover will reduce the probability of the AND term evaluating to 0, and hence reduces power consumption at the output of the AND term. At the same time, it will increase the probability of the output evaluating to 1, which increases the power consumption.

The example in Figure 4.5 shows how don't cares may be used to make power trade-offs at the output of the AND plane and the OR plane. Assuming each variable has 0.5 signal probability and $V_{dd}I_{dc(AND)}=1$, the output power cost of implicant P_1 (given by equation 4.1) is 0.75 while that of implicant P_{11} is 0.875. Therefore, output power cost is reduced when P_1 is used instead of P_{11}. A Similar statement holds for P_2 and P_{21}. However, if P_1 and P_2 are used instead of P_{11} and P_{12}, the output power cost at the output of the OR plane will be increased from 0.25 to 0.375.

PLA is a regular structure and the transistor sizes for the OR plane and the AND plane are identical. Therefore, $I_{dc(AND)}$ and $I_{dc(OR)}$ are the same. Based on this observation, the following theorem shows that including don't cares in the cover of a single

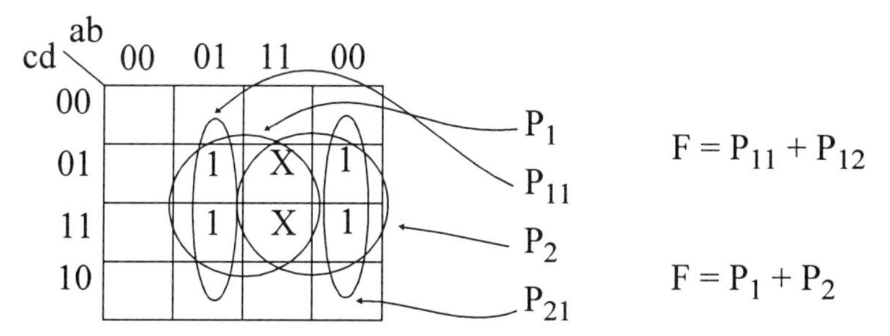

Figure 4.6 Including Don't Cares in the Cover

output function, in the worst case, will keep the power of the PLA unchanged, while in most cases, it will reduce the power consumption.

Theorem 4.1 *Including don't cares in the logic cover of a single output Boolean function will not increase the power cost of the final implementation of a pseudo-NMOS PLA.*

Proof: Let P be the set of minterms covered by the product term AND, and D be the subset of don't care minterms in P. When we subtract D from P, in the best case, the number of product terms remains unchanged. The minimum increase in power cost (due to the increase in sp_{AND}^{0}) is given by:

$$Power_{inc} = V_{dd} \cdot I_{dc} \cdot \sum_{dc \in D} sp_{dc}^{1} \tag{4.6}$$

The best reduction in power cost at the output when D is removed from the cover, occurs when D is only covered by p but no other product term. Hence, the maximum decrease in power is given by:

$$Power_{red} = V_{dd} \cdot I_{dc} \cdot \sum_{dc \in D} sp_{dc}^{1} \tag{4.7}$$

If some of the minterms in D are covered by other product terms in the AND plane, P_{red} will be smaller. From equations 4.6 and 4.7, removing don't care from the cover cannot increase the power cost.

■

Note also that when don't cares are included in a product term, in general, the number of inputs to the product term is also reduced. This will, in turn, reduce the power due to the input literals. The preceding analysis shows that without considering the power at the input, including don't cares, at worst, will not increase the power. Considering the reduction in power on the inputs, including don't cares will always result in reduction in power for a single output function.

The statement of theorem 4.1 does not, in general, hold for incompletely specified multiple output functions. This is because the power cost of a product term in the AND plane of a pseudo-NMOS PLA is independent of the load at its output. Consider a product term p and OR-terms $(o_1,..., o_n)$ that p fans out to. Removing don't care points from p will result in an increase in the power cost of p. Note, however, that this increase is independent of the number of fanouts. Now, removing don't care from p will in turn result in a decrease in the power cost of OR-plane terms $(o_1,..., o_n)$ if the don't cares are true don't cares for more than one OR-terms, and, at the same time, are not covered by other product terms. In general, it is possible for the reduction in power at the OR plane to be larger than the increase in power of p and, therefore, reducing the overall power cost. Approaches for dealing with multiple output functions are discussed in section 4.3.2.

The following theorem shows that power cost of a product term in a single output Boolean function is no more than any product term which it contains.

Theorem 4.2*Let power cost of a product term P for a single output Boolean function be given as:*

$$V_{dd} I_{dc} s p_P^0 \tag{4.8}$$

Let P_1 be a product term that is contained in P, then Power(P) $\leq$ Power(P_1).

Proof: Let M and M_1 be the sets of minterms in P and P_1 respectively. M_1 is a subset of M and let D be the difference of them. If D does not contain any don't care conditions, then in any case, D must be included in the function output and hence including D in the prime implicant P will not increase power cost at the NOR gate of the output plane. Thus we have:

$$\begin{aligned} PowerCost(P) - PowerCost(P_1) &= V_{dd} \cdot I_{dc} \cdot \left(\sum_{i \in M} sp_i^0 - \sum_{j \in M_1} sp_j^0 \right) \\ &= V_{dd} \cdot I_{dc} \cdot \left(\sum_{j \in M_1} sp_j^1 - \sum_{i \in M} sp_i^1 \right) \\ &= -V_{dd} \cdot I_{dc} \cdot \sum_{k \in D} sp_k^1 \leq 0 \end{aligned} \tag{4.9}$$

In the case that D includes don't care conditions, let D_1 be the sets of don't care conditions in D. From theorem 4.1, including don't care conditions will not increase the power cost of a product term. Therefore, in the worst case equation 4.9 becomes:

$$PowerCost(P) - PowerCost(P_1) = -V_{dd} \cdot I_{dc} \cdot \sum_{k \in D_s} sp_k^1 \tag{4.10}$$

where D_s is the difference of D and D_1 and D_1 is only covered by P. In the worst case, D_1 is equal to D and equation 4.10 is equal to zero.

■

4.2.2 Dynamic-CMOS PLA

Assuming that NOR gates in the AND and OR planes drive the same capacitive loading, the effect of using don't care conditions in the logic cover will be the same as the case for pseudo-NMOS PLAs. Also note that for dynamic PLA implementations, this result also holds for multiple output functions. This is easily proved by noting that AND plane power for dynamic CMOS PLAs is a function of its load. This load is proportional to the number of OR-terms that this product term fans out to. Removing don't cares from a product term p will result in even more increase in power as the number of fanouts of p increases. This is in contrast with pseudo-NMOS circuits where the power in the AND plane is independent of its load. Consider equation 4.4 for the power cost of a product term for dynamic PLAs. The final term in this equation ($2*c_{clock}$) is a constant term and is dropped in the optimization.

Theorem 4.3*Let Power cost of a product term P be given as:*

$$\frac{V_{dd}^2 \cdot f}{2} \cdot \left(\sum_{i=1}^{k} c_i sw_i + c_p sp_p^0 \right) \tag{4.11}$$

Let P_1 be a product term that is contained in P, then Power(P) $\leq$ Power(P_1).

Proof: Let p and p_1 be the set of literals used in P and P_1. p is a subset of p_1 since P contains P_1. Therefore:

$$\sum_{x_i \in p} c_i \cdot sw_{x_i} - \sum_{x_j \in p_1} c_j \cdot sw_j \leq 0 \tag{4.12}$$

Also from equation 4.11, it is shown that:

$$sp_P^0 \leq sp_{P_1}^0 \tag{4.13}$$

If we replace P by P_1 in a cover, the number of product terms in the sum of product form in the logic cover will, at the best case, unchanged. Therefore:

$$C_P \leq C_{P_1} \tag{4.14}$$

Combining equations 4.11, 4.12 and 4.13 proves the claim of this theorem.

■

This theorem shows that only prime implicants need to be considered in searching for the minimum power solution in dynamic-CMOS PLA optimization.

4.3 Two Level Function Minimization

Traditionally, two level logic minimization for PLA targets minimum area. PLA is a regular structure, and its area is proportional to (number of inputs + number of outputs) × (number of product terms). Therefore, the problem of minimizing the PLA area is equivalent to the problem of finding the minimum number of product terms to implement the Boolean function. For power minimization, the objective is to minimize the power consumption at both the AND plane and the OR plane. The following sections show that the relationship between optimizing area and optimizing power for PLA minimization, and then describes how the algorithms for area minimization are modified to minimize power consumption.

4.3.1 Area optimization

As stated before, minimizing area in a PLA is equivalent to minimizing the number of product terms. The PLA area is relatively independent of the implementation technology. Therefore, the minimization algorithms that target minimum area work equally well for different types of technologies. It is easy to show that if the cost of a product term is not larger than the cost of any product term which it contains, then a minimum cost solution consists only of prime implicants. For PLA area minimization, the cost of a product term is equal to the area for adding a row to the PLA, and is constant for every product term. Therefore the above condition is satisfied and only prime implicants need to be considered during the minimization process. The classic solution for finding a minimum cover for a function is due to Quine-McCluskey. This technique consists of the following steps:

1: Generate the set of all prime implicants.
2: Form the prime implicant table.
3: Derive a minimum cover for this table.

For generating the set of all prime implicants, some algorithms resort to explicit enumeration of the minterms which, in worst case, is exponential in complexity. Even if the prime implicants are derived without enumerating minterms, there exist functions with an exponential number of prime implicants as a function of the number of implicants in a minimum cover. Also, the minimum-cover problem itself is NP-complete.

Rudell [3] proposed an exact algorithm to solve the two-level minimization problem based on the motivation that the problems faced in reality do not have worst case behavior. An exact solution can also be used to measure the quality of heuristic methods. Heuristic methods such as ESPRESSO [1] were proposed to solve problems with large number of inputs and product terms. ESPRESSO uses an iterative improvement to achieve confidence in the optimality of the final result. Instead of generating all prime implicants, and finding a minimum cost cover, implicants are expanded to prime implicants and then an irredundant cover is generated from the expanded prime implicants. This approach eliminates the basic problem of explicit generation of all prime implicants. To avoid local minima, after the expansion and removal of covered implicants, the remaining implicants are maximally reduced while still maintaining a cover. Then the expansion process is repeated, and the entire procedure is iterated until no improvement is obtained. The sequence of operations carried out by Espresso is as follows:

1: Compute the Off-set and the DC-set.

2: *Expand each implicant into a prime implicant and remove redundant primes. During the expansion process, "blocking matrix" and "covering matrix" are formed to guide the choosing of the lowering set and the raising set which are the sets of literals that are going to be lowered and raised in the expanded prime implicants respectively.*
3: *Extract the essential primes and place them in the DC-set.*
4: *Find an irredundant cover such that the cardinality of the cover is minimum.*
5: *Reduce each prime implicant in the irredundant cover to a minimum essential implicant.*
6: *Iterate steps 2, 4, and 5 until no further improvement is made.*
7: *Try steps 5, 2, and 4 one last time using a different strategy. If successful, continue the iteration.*
8: *Include the essential primes back into the cover and make the PLA structure as sparse as possible.*

4.3.2 Power optimization

As shown in this chapter, only prime implicants need to be considered for completely specified single output functions, and incompletely specified multiple-output functions which target dynamic CMOS and pseudo-NMOS implementations. In these cases, Quine-McCluskey procedure is used to find the minimum power solution. In building the prime implicant table, each column representing a prime implicant is tagged with the power cost given by equation 4.1 or 4.4.

For an exact solution, the algorithm described in Espresso-Exact [3] is modified by using appropriate weight functions during the branch-and-bound process that is used to find a minimum cost cover. Instead of using the number of prime implicants in the partial solution as the current cost, the sum of the power costs of the corresponding prime implicants is used. In case of incompletely specified single output function, power cost incurred at the output of the functions should also be added to the total cost. This additional power is caused by including don't cares in the cover. The power cost of a partial cover is then given by:

$$\sum_{p \in PartialCover} PowerCost(p) + \sum_{d \in DC_{PartialCover}} V_{dd} \cdot I_{dc} \cdot sp_d^1 \tag{4.15}$$

where $DC_{partialCover}$ is the set of don't care that is included in the partial cover. $DC_{PartialCover}$ can be found by augmenting the prime implicant table by adding a row for each element in the *DC_set* of the function. If a row is covered by one of the prime implicants already in the partial cover, then the corresponding don't care point is in $DC_{PartialCover}$.

For incompletely specified multiple output functions that target pseudo NMOS PLAs, it is possible to obtain a lower power implementation by removing don't cares from product terms. We define a *pseudo prime implicant* as follows:

Definition 4.1 *Given a Boolean function f and its don't care set, a pseudo-prime implicant is defined as an implicant p such that if it is possible to expand p in any direction, then the new points in the on-set of p will only include don't care points.*

Note that for a function with no don't care, the set of pseudo-prime implicants are identically the same as the set of prime implicants of the function. Also note that each pseudo-prime implicant P_p corresponds to a prime implicant P of the function where P_p and P contain the same on-set points of the function. The incompletely specified multiple output pseudo-NMOS problem is then solved by using the minimum covering problem to select the set of pseudo-prime implicants, which results in minimum total power cost. Once a minimum power solution is obtained, each pseudo-prime implicant in the cover is expanded after checking to make sure this expansion does not increase the total power. Note that this approach does not guarantee an exact minimum solution. Another approach for solving the problem for incompletely specified multiple-output pseudo-NMOS PLA is to solve the exact covering problem using the prime implicants of the function, and then reduce each prime implicant in the final cover if this reduction results in a decrease in total power. Special cases may also be considered during this step. For example theorem 4.1 states that if a don't care set is included in only one output function, then that don't care set may freely be included in the on-set.

For a heuristic solution, Espresso can be used with the following modification. Let $C(F)$ be a cover of the function F. If every implicant in $C(F)$ is used in the final implementation, the power cost is given by:

$$\sum_{I \in C(F)} PowerCost(I) \tag{4.16}$$

If an implicant is expanded to a prime implicant PI which covers implicants in $C_1 \subset C(F)$, then including PI in the cover will reduce the power cost by:

$$\sum_{I \in C_1} PowerCost(I) - PowerCost(PI) - \sum_{dc \in DC} PowerCost(dc) \tag{4.17}$$

where DC is the set of additional don't care introduced in the output when PI is included in the cover. During the expansion step, each implicant is expanded to a prime such that the reduction in power cost specified in equation 4.17 is maximized.

While finding an irredundant cover, the cover $C(F)$ that minimizes the following cost function is generated:

$$\sum_{I \in C} PowerCost(I) + \sum_{dc \in DC} PowerCost(dc) \tag{4.18}$$

For the reduction step, the prime implicant is reduced such that after reduction, the cover $C(F)$ has minimum power cost given by equation 4.18.

4.4 Experimental Results

In order to measure the effectiveness of power optimization in PLAs, the procedure for ESPRESSO-exact was modified to obtain low power implementations of pseudo-NMOS and dynamic CMOS PLAs. Table 4.1 gives the results of these experiments for dynamic CMOS PLAs. Columns 1, 2, and 3 give the number of product terms (cubes), number of literals in the SOP form, and the power of the resulting implementation when the examples are minimized for minimum area. Columns 4, 5, and 6 give the same values when the circuits are optimized for low power. Note that the values in the last three columns are normalized with respect to their corresponding values in the first three columns. It can be seen that optimization for power has resulted in a 5% increase in the number of cubes of the function, while reducing the power by an average of 11%. In general, optimizing the functions for power has also resulted in functions with fewer literals in the SOP form. For pseudo-NMOS PLAs, the power reduction is not as significant as that of dynamic CMOS PLA. The reason is that the power due to input literals is small compared to the static power drawn in the NOR gates. Therefore, the effect of the number of product term used on power consumption is more significant than the input literal power. The minimum area solution which gives a minimum number of product terms will hence provide a good solution for low power. The results of the exact solutions show that, in general, a minimum literal PLA also provides a good low power solution. Based on the results of exact solutions, it is not expected for heuristic solutions for low power to provide significant improvements in power over the current minimal area solutions obtained using programs such as ESPRESSO.

	Minimum Area Solution			Minimum Power Solution		
ex	cubes	literals	power	cubes	literals	power
Z5xp1	63	419	175.74	1.03	0.90	0.82
amd	66	738	348.87	1.03	0.98	0.96
b10	100	1220	485.24	1.02	0.98	0.95
dekoder	9	57	28.43	1.11	0.91	0.82
dk48	21	301	56.78	1.14	0.40	0.59
f51m	76	447	164.17	1.00	0.98	0.96
gary	107	1226	431.98	1.02	0.99	0.96
in0	107	1226	446.42	1.02	0.99	0.96
in2	134	1532	472.58	1.02	0.98	0.96
inc	29	241	116.50	1.07	0.93	0.89
log8mod	38	243	86.23	1.08	1.00	0.95
m181	41	288	129.65	1.05	0.86	0.81
newcpla1	38	350	168.19	1.05	0.97	0.95
opa	77	1357	864.90	1.06	0.97	0.94
root	57	412	142.84	1.00	0.95	0.91
sqr6	47	331	155.79	1.09	0.91	0.83
wim	9	60	37.08	1.11	0.92	0.90
AVG				1.05	0.92	0.89

Table 4.1 Power reduction in PLA optimization

References

[1] R. Brayton, G.D. Hachtel, C. McMullen and A. Sangiovanni-Vincentelli. "Logic Minimization Algorithms for VLSI Synthesis." *Kluwer Academic Publishers*, Boston, 1984.

[2] S. Devadas, K. Keutzer and J. White. "Estimation of Power Dissipation in CMOS Combinational Circuits using Boolean Manipulations." In *IEEE Transactions on Computer Aided Design*, vol 11, no 3, March 1992.

[3] R. Rudell. "Logic Synthesis for VLSI Design." Ph.D. thesis, University of California, Berkeley, 1989.

Part III

Multi-level Network Optimization for Low Power

CHAPTER 5 Logic Restructuring for Low Power

Multi-level implementations of digital circuits are in general smaller and consume less power compared to two-level implementations of the same logic. In fact, in some cases where a two-level implementation has exponential size in the number of inputs (i.e. a multi-input XOR function), the multi-level implementation provides a non-exponential implementation. Logic restructuring techniques are an important part of logic synthesis as they allow for creation of multi-level designs from two-level representation of the circuit. Logic restructuring techniques are also used in optimizing the multi-level representation of circuits. Logic restructuring techniques include common sub-function extraction and factorization. Common sub-function extraction for a set of logic functions in a Boolean network is the process of determining an optimal set of logic functions such that when inserted into the network, the cost of the network is minimized. Factorization is the process of writing the logic function in a parenthesized form such that the target cost of the network is minimized. The complexity of finding the factorized form and decomposition of nodes is very high for any real sized network. Therefore approximations need to be made in order to make the problem tractable. A powerful approach introduced in [2] is to assume that logic functions in the network are algebraic expressions. Using this approach, only common sub-expressions instead of sub-functions, are considered for extraction and factorization. The approach is to first make all functions prime and irredundant, and then look for sets of cubes which divide two or more expressions in a formal, algebraic way. Furthermore, by restricting computations to formal manipulations, fast algorithms for division and other similar operations are obtained. These operations include algebraic based extraction algorithms which provide fast methods for identifying logic sharing among different nodes, substitution algorithms which help in reducing the area by taking advantage of functions that have already been implemented, algebraic factorization algorithms which provide fast techniques for estimating the area of the network by computing the number of literals in the factored form of the network, and also decomposing nodes into smaller nodes by taking advantage of logic sharing within the same node. The application of these procedures during technology independent phase of logic synthesis has proved to be quite effective. The formulation of *Rectangle Covering Problem* introduced in [3] provides an efficient technique for generating the set of kernels, kernel intersections, and common cubes [2] which represent good candidate divisors of algebraic expressions.

The traditional cost function using algebraic techniques has been network area. This, in general, has been represented by the number of literals in the network. During algebraic operations, the value (or effectiveness) of a divisor is measured in terms of the reduction in the number of literals that is obtained by using that divisor. This chapter describes algebraic procedures for node extraction and factorizations that target minimum power consumption in the network. The power cost functions introduced in section 2.4 are used to measure the quality of candidate divisors in the network.

5.1 Algebraic Logic Restructuring

Algebraic techniques were proposed to reduce the complexity of extraction and decomposition of Boolean functions. The methods described here use algebraic techniques to perform extraction and decomposition targeting low power. In the following, some basic definitions for algebraic techniques are provided.

A product $f.g$ is an *algebraic product* if f and g have disjoint supports. Given two expressions f and g, a *division operation* can be defined which computes q and r such that $f=q.g+r$. The division is algebraic if the product $q.g$ is an algebraic product. f/g denotes algebraic division of expression f by expression g. An expression is called *cube-free* if no cube divides the expression evenly. The *primary divisors* of a expression f, $D(f)$ are the expressions f/c where c is a cube. The *kernels* of an expression are then defined as the cube-free primary divisors of f. The set of all kernels of f is written as $K(f)$. The cube c used to generate kernel $k=f/c$ is called the *co-kernel* of f [2]. *Weak division* is used to perform algebraic division [9].

5.2 Common Sub-Expression Extraction

Algebraic extraction uses the concepts of common kernel and cube extraction to introduce new nodes into a network in order to optimize the area cost of a Boolean network. These techniques use the SOP cost of nodes to minimize the total number of literals in the SOP form of the network. Algebraic extraction techniques utilizing kernels and cubes are used extensively, and result in significant reduction in the area of the technology mapped network [6].

The problem of power optimization during extraction is stated as follows:

Problem: *Given a boolean network, find a set of nodes to introduce into the network such that the power cost of the network in the SOP form is minimized. The power cost of the network in the SOP form is the sum of the SOP power cost of all nodes in the network.*

Kernel and cube extractions can be used to minimize the network power consumption. These techniques are discussed in the following sections.

5.2.1 Cube and kernel extraction

During area optimization, the value of a kernel is defined as the reduction in the number of SOP literals in the network after the kernel is extracted. The following example illustrates this computation:

Example 5.1:

$$F = a\,b + b\,c + a\,c$$

$$G = a\,d + b\,d + c\,d$$

All kernels of the functions are:

$K_1 = a + b$	co-kernel: c
$K_2 = a + c$	co-kernel: b
$K_3 = b + c$	co-kernel: a
$K_4 = a + b + c$	co-kernel: d

Kernels K_1, K_2 *and* K_3 are three possible kernel intersections that can be extracted out of both functions. Extracting K_3 will result in the following set of equations:

$$F = a\,K_3 + b\,c$$

$$G = d K_3 + a d$$
$$K_3 = b + c$$

By extracting K_3 total number of literals in the SOP form is reduced by two literals. Hence the value for this kernel intersection is 2. Kernels K_1 and K_2 also have a value of 2. This means that during extraction, maximum gain is obtained by extracting either K_1, K_2 *or* K_3.

As shown in the previous example, kernel extraction for area may select any of three kernels for extraction. However, the same is not true when power is being optimized. In general, it is also possible for a kernel K_1 with a higher area value than K_2 to have a lower power value than K_2 as shown in the following example.

Example 5.2:

Consider the following function: $F = a\,b\,c\,d\,e + d\,e\,f\,g\,h\,i + f\,g\,i\,j\,k$.

Also assume $sp(f)=0.1$ and signal probability of all other fanins is 0.5.

All kernels of function F are:

$K_1 = a\,b\,c + f\,g\,h\,i$ Co-kernel: d

$K_2 = d\,e\,h + j\,k$ Co-kernel: $e\,f\,h$

Function F can be represented as :

$$F_1 = K_1\,d\,e + f\,g\,i\,j\,k$$
$$F_2 = a\,b\,c\,d\,e + K_2\,f\,g\,i$$

Assuming an SOP implementation, extracting K_1 will reduce the number of literals by 1 and the power by 0.523. Extracting K_2 will reduce the number of literals by 2 and power by 0.137.

These examples show that in order to minimize the power consumption of a network, it is necessary to look at the power cost of the function. This also means that a minimum area solution is not necessarily the minimum power solution.

5.2.2 Kernel extraction targeting low power

In [5] a modification of kernel extraction algorithm is presented which generates multi-level circuits with low power consumption. The procedure follows.

Let $d=d(v_1,...v_M)$, $M>0$ be a common sub-expression of functions $f_1,...f_L$, $L>1$. Let $(J_1,...J_P)$, $P>0$ be the nodes internal to d. When d is factored out of f_i, the signal probabilities and switching activity at all nodes of the network remain unchanged.

However, the load at the output of the driver gates $(v_1,..v_M)$ change. Each gate now drives L-1 fewer gates. At the same time, since there is only one copy of d instead of L copies, there are L-1 fewer copies of internal nodes $J_1,..J_P$. The power saving in extracting d is thus given as:

$$(L-1)\cdot\left(\sum_{(i=1)}^{M} sw(v_i)\cdot n_{v_i} + \sum_{(i=1)}^{P} sw(J_i)\cdot n_{J_i}\right) \tag{5.1}$$

where n_{v_i} gives the number of gates belonging to d and driven by signal v_i and n_{J_i} is the number of gates internal to d and driven by signal J_i. One shortcoming of this method is that first a factored form for the functions is assumed and the sub-expressions are extracted based on these factored representation. However, once these sub-expressions are extracted, they will not necessarily have the assumed factorized form. This introduces an inconsistency in the flow of the procedure which will potentially lead to inconsistent results.

An alternative approach for computing the power value of extracting a common sub-expression from a set of logic functions is described below. This power value uses the power cost of a node in the SOP form to compute the value for extraction. Using power cost in SOP form is consistent with the assumption made for computing the literal-saving value of a candidate divisor during algebraic extraction. Consider a multiple-output Boolean function $F=\{f_1,..f_L\}$ with cubes $(c_1,...,c_N)$ and input set $(v_1,...v_M)$. Let $D=(d_1 +.. + d_P)$ represent a kernel of function f used as a divisor. Also assume $Q=\{q_1,..q_R\}$ is the set of co-kernels for kernel D in functions $\{f_1, ..,f_L\}$. The area value for extracting D is given by equation 5.2 [6]. In this equation, the first term accounts for literals saved by not repeating the kernel, the second term accounts for literals saved by not repeating the co-kernels, and the last term accounts for the number of literals introduced by extracting kernel D.

$$\left\{(R-1)\cdot\sum_{(i=1)}^{P} lit(d_i)\right\} + \left\{(P-1)\cdot\sum_{(i=1)}^{R} lit(q_i)\right\} - R \tag{5.2}$$

The *power_value* for extracting kernel D is defined in equation 5.3. In this equation, the first term accounts for the load reduction on the inputs to the kernel. The second term accounts for the load reduction on the inputs to the co-kernels of the given kernel. The third term accounts for the cubes of the function that are removed from the original SOP representation of the function. Note that in this representation $(d_i.q_j)$ corresponds to a cube of the original SOP representation of the function. The forth term corresponds to power consumption at the output of the new node which is inserted into the network. The fifth term corresponds to the power consumption at the output of the cubes of the new node that is inserted into the network. The last term accounts for the power at the output of the new cubes inserted in the SOP representation of function *f*.

$$\left((R-1)\cdot\sum_{(i=1)}^{M} sw(v_i)\cdot CL(v_i, D)\right)+\left((P-1)\cdot\sum_{(i=1)}^{R}\sum_{(j=1)}^{M} sw(v_j)\cdot CL(v_j, q_i)\right)$$
$$+\left(\sum_{(i=1)}^{P}\sum_{(j=1)}^{R} sw(d_iq_j)\right)-(R\cdot sw(D)) \tag{5.3}$$
$$-\left(\sum_{(i=1)}^{P} sw(d_i)\right)-\left(\sum_{(j=1)}^{R} sw(Dq_j)\right)$$

The following example uses these equations to compute the power and area value for extracting a kernel.

Example 5.3:

Assume:

$$F_1 = a\,x\,y + a\,u\,w + v\,z$$
$$F_2 = b\,c\,x\,y + b\,c\,u\,w + v\,z$$

and $D = x\,y + u\,w \quad \Rightarrow q_1 = a,\ q_2 = bc \text{ and } R = P = 2$

Also assume the following signal probabilities for the circuit inputs:

$sp(a) = 0.97$ $\quad sp(u) = 0.91$ $\quad sp(x) = 0.67$

$sp(b) = 0.02$ $\quad sp(v) = 0.93$ $\quad sp(y) = 0.47$

$sp(c) = 0.51$ $\quad sp(w) = 0.35$ $\quad sp(z) = 0.65$

Using the given values, we compute the switching activity for the following product terms and functions:

$sw(axy) = 0.424$ $\quad sw(auw) = 0.427 sw(vz) = 0.478$

$sw(bcxy) = 0.006$ $\quad sw(bcuw) = 0.006$

$sw(aD) = 0.500$ $\quad sw(bcD) = 0.011$

$sw(xy) = 0.431$ $\quad sw(uw) = 0.439$

$$sw(D) = 0.498 \quad sw(F_1) = 0.309 \quad sw(F_2) = 0.011$$

If D is extracted from F_1 and F_2, then

$$F_1 = a\,D + v\,z, \qquad F_2 = b\,c\,D$$

$$D = x\,y + u\,w$$

Using equation 5.2, the area value for kernel D is computed as:

$$(2-1)*4 + (2-1)*(1+2) - 2 = 5$$

In fact the number of literals of the functions is reduced by 5 after D is extracted.

Using equation 5.3, the power value for this extraction is computed as follows:

$$\begin{aligned}
&(2-1) * (sw(x) + sw(y) + sw(u) + sw(w)) \\
&+ (2-1) * (sw(a) + sw(b) + sw(c)) \\
&+ (sw(axy) + sw(auw) + sw(bcxy) + sw(bcuw)) \\
&- 2 * sw(D) \\
&- (sw(xy) + sw(uw)) \\
&- (sw(aD) + sw(bcD))
\end{aligned}$$

where each line in this equation corresponds to one term in equation 5.3. Then

power_value = $1.559+0.597+0.863-0.996-0.865-0.511$

$= 0.647$

Figure 5.2 shows the procedure for performing low power kernel extraction for a Boolean network. In this procedure, the set of all kernels for all the nodes in the network is first computed using the procedure described in [6]. Procedure *best_power_subkernel* returns a kernel intersection of kernels in K which has maximum power value. The sub-expression corresponding to this maximum value is then extracted. The list of all network kernels is then updated by including the kernels for the newly created node. This operation is performed as long as the power value is larger than a user defined threshold value.

Power tracks well with the area of a network. Therefore, it is desired that power optimization techniques do not degrade the area quality of the solution while optimizing the network for power. The following lemma states a relationship between the power value and area value for an extraction.

Lemma 5.1 *Given a set of functions $F=\{f_1, .., f_L\}$, and sub-kernel D with co-kernels $Q=\{q_1,..q_R\}$ to be extracted from F, if power value of D is greater than or equal to zero then area value of D is greater than or equal to zero.*

```
1: function power_kernel_extract(G)
2:   G(V, E) is a Boolean network;
3:   begin
4:     do
5:         K = 0; value = 0
6:         foreach node n ∈ G do
7:             K = K U (the set of all kernels for node n)
8:         endfor
9:         K_i = best_power_subkernel(K_i)
10:        extract K_i from nodes having K_i as a sub-expression.
11:        K = K U (set of all kernels for K_i)
12:        value = power_kernel_value(K_i)
13:    while( value > thresh)
14:  end
```

Figure 5.1 kernel extraction for low power

Proof: Assume that the number of literals in kernel D is given by K and number of literals in co-kernels of D is C. Then the area value of D can be written as $(R-1)(K-1)-1+(P-1)C$. Considering that all variables in this equation have to be greater than zero, solving this linear integer equation shows that it has a negative value only when $K=1$ or when $R=1$ *and* $P=1$. If $K=1$ then kernel D has only one literal $D=l$ and we must have $P=1$. Under these conditions, the power value of kernel D will be $-2.sw(l)$. The power value evaluates to $-2.sw(d)$ in the second case where d is the single cube in kernel D. We see that conditions implied by a negative area value for kernel D imply a negative power value for this kernel. This implies the claim of the lemma.

■

This lemma states that extractions performed using *power_kernel_extract* will not increase the literal count in the network. This is an important property since if the area of the circuit is increased significantly during power extraction, then the power due to the increased area will more than offset the reduction in power and no power gain will be obtained.

Our experiments show that in general the sequence of sub-expressions extracted by *power_kernel_extract* also has a decreasing area value and that the power value is mainly used to break ties between sub-expressions which have equal area value. It is possible for a sub-expression with a low area value to have a high power value. In this case, this sub-expression will be extracted before other sub-expressions. The majority of extractions, however, follow a decreasing value in area, as well as a decreasing value in power.

5.2.3 Cube Extraction Targeting Low Power

Consider a single-output Boolean function f with cubes $(c_1,...,c_N)$ and input set $(v_1,...v_M)$. Let $D = c_i \cap c_j$ be a sub-cube of the function with T literals that is shared by $R(>1)$ cubes of the function. The area value for extracting D is given by:

$$(R-1)\cdot(T-1)-1 \tag{5.4}$$

The power value for extracting cube D is then defined by equation 5.5. The first term in this equation accounts for the load reduction on the inputs fanning out to D. The (T-1) term accounts for the power introduced in the network by adding a gate into the network. The load for this new gate is $CL(f, D)$ which in this case is equal to R.

$$(R-1)\cdot\sum_{(i=1)}^{M} sw(v_i)\cdot CL(v_i, D) - R\cdot sw(D) \tag{5.5}$$

The power value presented here can easily be extended to multiple functions by considering cubes of all functions while searching for cube intersections.

Figure 5.2 shows the procedure for performing power cube extraction for a Boolean network. Procedure *best_subcube* looks at all intersections of two cubes in the network and finds the cube with maximum power value. The sub-cube with maximum value is introduced into the network. This operation is repeated as long as there exists a sub-cube with a positive power value.

The following lemma shows that a cube extraction that leads to a power reduction, will not result in an increase in the area cost of the network. As stated before, this is an important property since it is important to provide a low area solution while generating a low power solution.

Lemma 5.2 *Given a set of functions $F=\{f_1, ..,f_L\}$, and cube D with T literals to be extracted from F, if the power value of D is greater than or equal to zero then area value of D is greater than or equal to zero.*

Proof: If the area value of D is less than zero we must have either T=1 or R=1 in

```
1: function power_cube_extract(G)
2:   G(V, E) is a Boolean network;
3:   begin
4:      do
5:          D= best_subcube(G)
6:          extract D from all nodes having D as a sub-cube
7:          value += power_cube_value(D)
8:      while( value > thresh)
```

Figure 5.2 Cube extraction for low power

equation 5.4. For both conditions the power value of D evaluates to $-sw(D)$. This means that a negative area value implies a negative power value, which proves the lemma.

■

This lemma also demonstrates that any cube extracted using *power_cube_extract* does not increase the area of the network. As with sub-kernel extraction, experiments show that most of the time, the sequence of cubes extracted using *power_cube_extract* also has a decreasing area value.

5.2.4 Quick power extract

The procedures shown in figures 5.1 and 5.2 require that a power value be computed for all sub-expressions of the network. The shortcoming of the power value as presented in equation 5.3 and 5.5 is that for each sub-expression being extracted, the switching activity of all the cubes being removed and being inserted need to be computed.

Correct computation of the switching activity for any internal node or function (under a zero delay model) [4][7] requires that the global OBDD and then the signal probability and switching activity of the function be computed. This operation proves to be very time consuming. In order to speed up the extraction procedure, the signal probability and switching activity values on immediate fanins of the node are used to approximately compute its switching activity.

5.3 Function Factorization

5.3.1 Maximally factored logic expressions

Factorization is the process of deriving a factored form from a sum-of-products form of a function. For example, if $F=a.b+a.c+b.c$ then one possible factorization of F is $a.(b+c)+b.c$. In most logic synthesis systems Boolean functions are internally stored in the sum-of-products form. The area of a Boolean network, however, is more accurately estimated by the number of literals in the factored form of the network. This means that an efficient factoring algorithm is needed in order to guide the optimization procedures. Exact methods for computing the best factorized form of a function have been presented in the past. However, these algorithms are too complex to be used since the factorization algorithms are used to guide the optimization procedure and have to be performed numerous times. In this section, a heuristic technique presented in [9] is used to compute the factorized form of a Boolean function. The factored form of a node is then used to compute the area and power cost of a node.

A heuristic factoring algorithm is described in [9]. In this algorithm a recursive procedure is used to find a factored form of a given sum-of-products representation. At each step of the recursion, the function F passed to the procedure is transformed into $F=Q.D+R$ where D is the divisor, Q is the quotient and R is the remainder of dividing F by D. The procedure guarantees that the resulting factorized form of F is maximally factorized. The procedure *DIVISOR* passed to the function is used to find a candidate divisor for the function. By changing this procedure, trade-offs in terms of speed and quality of results can be made. Quotient Q is computed by performing weak division on F and D.

Procedure *literal_factor* (figure 5.2) is a variation of literal factoring algorithm. *best_literal* selects a literal in C which occurs in the largest number of cubes of F. *Common_cube* returns the largest common cube.

5.3.2 Factorization for low power

The procedure *generic_factor* as presented in 5.3.1 guarantees that the algebraic expression is maximally factorized for the choice of the divisors, provided at each step of the recursion. A maximal factored form will ensure that for the given choice of divisors, each literal is used a minimum number of times. In fact, the quality of the results is determined by the value of the divisor which is returned by the *DIVISOR*

```
1:  function generic_factor(F, C, DIVISOR)
2:    F is an algebraic expression, C is a cube
3:    DIVISOR  a procedure to find a divisor
4:    begin
5:      if ( (D= DIVISOR(F)) ) == 0 ) then
6:          return F
7:      endif
8:      Q = WEAK_DIVIDE(F, D)
9:      if ( |Q| = 1 ) then
10:         return literal_factor(F, Q, DIVISOR)
11:     else
12:         Q = make_cube_free(Q)
13:         (D,R) = WEAK_DIVIDE(F, Q)
14:         if ( cube_free(D) ) then
15:             Q = generic_factor(Q, DIVISOR)
16:             D = generic_factor(D, DIVISOR)
17:             R = generic_factor(R, DIVISOR)
18:             return Q * D + R
19:         else
20:             C = common_cube(D)
21:             return literal_factor(F, C, DIVISOR)
22:         endif
23:     endif
24: end
```

Figure 5.3 Maximal factoring of logic expressions

procedure. In addition to *DIVISOR*, procedure *best_literal* in (figure 5.2) is the only step in the factorizing procedure which affects the quality of the factorized form. All other steps of the factorizing procedure are necessary to guarantee a maximally factored expression.

The goal of the factorizing procedure is to obtain a factored form with a minimum number of literals. In order to achieve this goal, the heuristic presented here attempts to maximally reduce the number of literals in the factored form at each step of the recursion. This means that given an algebraic expression F, this procedure will return a partially factored expression $Q.D+R$ with the minimum number of literals. The choice of the divisor is intended to result in maximal reduction in literals at each step of the recursion. In fact, procedure *best_literal* is used to find a single literal divi-

```
1: function literal_factor(F, C, DIVISOR)
2:   F is an algebraic expression, C is a cube
3:   DIVISOR a procedure to find a divisor
4:    begin
5:        L = best_literal(F, C)
6:        (Q, R) = DIVIDE(F, L)
7:        C = common_cube(Q)
8:        Q = generic_fator(Q, DIVISOR)
9:        R = generic_factor(R, DIVISOR)
10:       return L * C * Q + R
11:   end
```

Figure 5.4 Dividing expressions by literals

sor for expression *F*. Given an algebraic expression *F*, selecting literal *L* used in maximum number of cubes of *F* as divisor will result in maximal reduction in the number of literals of the expression *Q.L+R.*

The problem of factorization for low power is stated as follows:

Problem: *Given a sum-of-products expression for function F, find a maximally factored expression where the weighted sum of the literals by their switching activity is minimized.*

The heuristic in figure 5.2 is used to ensure that the resulting factored forms are maximally factored. The procedures for finding a divisor, however, are modified to take into account the new cost function. For power extraction *best_power_literal* as shown in figure 5.5, is used to select a literal in procedure *literal_factoring.*

Lemma 5.3*Given an algebraic expression F, selecting literal l returned by best_power_literal will maximally reduce the weighted sum of the literals in the factored form of the expression Q.l+R obtained by dividing l into F.*

Proof: Reduction in the weighted sum of the literals is given by $(CL(l,F)-1)t(l)$. Maximally reducing the weighted sum of literals is equivalent to maximizing $(CL(l,F)-1)t(l)$, and this expression is maximized by maximizing $CL(l,F)*t(l)$.

■

The quality of factorization is directly affected by the divisor returned by the *DIVISOR* procedure. Quick and good factoring are performed by using *generic_factor*, and using two different methods for divisor selection. Quick factor-

```
9:  function best_power_literal(F, C)
10: F is an algebraic expression, C is a cube
11: begin
12:     L = L1
13:     foreach Li ∈ C do
14:         if( E(Li)*CL(Li, F) < E(L)*CL(L, F) ) then
15:             L = Li
16:         endif
17:     endfor
18:     return L
19: end
```

Figure 5.5 Finding best literal for power

ization is used as a very fast factorization technique. Quick factor is implemented by calling the *generic_factor* procedure with *quick_divisor* passed in as *DIVSOR* parameter. The procedure *quick_divisor* returns a level zero kernel of the expression F. The procedure is guided by selecting at each step of recursion, the literal which appears in most cubes of the expression. This choice of literal removes most frequently used literals of F from the divisor to be returned. This means that these literal will be candidates to be selected by the *best_literal* procedure called in *literal_factor*. This choice also will result on level zero kernels with more cubes which will, in turn, lead to more reduction in the number of literals of the factored form.

In order to extend the *quick_divisor* procedure to account for power, the *best_literal* procedure in *quick_divisor* is replaced by *best_power_literal*. This means that literals that are better candidates for power will be considered in *generic_factor* by calling *best_power_literal* in *literal_facor*.

In summary quick power factorization is implemented by replacing all occurrences of *best_literal* in the procedure *generic_factor*, and all procedures called by this function with *best_power_literal*.

Good factorization is obtained by calling the *generic_factor* procedure with a *good_divisor* procedure passed in as *DIVISOR* parameter. *good_divisor* will look at all kernels of the function, and selects a kernel with maximum area value, as defined in equation 5.2. This procedure is extended to minimize the power cost by selecting the kernel with maximum *power_value*, as defined in section 5.2.2.

Function Decomposition is performed by first finding an optimal factored form of the logic function. Factored forms are graphically represented as labeled trees, called factoring trees where each node is labeled as "+" or "*" and each leaf is labeled by a literal. Decomposition is performed by building the decomposed form from the factored tree by replacing each internal node by a gate corresponding to its label.

5.4 Expression Substitution

Substitution of a function G into F is the process of re-expressing F in terms of variable G, and the original inputs of the function F [9]. Substitution can be performed using algebraic or Boolean division algorithms. Even though Boolean division, in general, yields better results, algebraic division is a much faster heuristic which produces comparable results. If algebraic division is used for substituting variable G into function F, and the quotient of the result is not 0, then we can conclude that G was a sub-expression of function F. In fact, if a non-trivial expression G (a function other than buffers or wires) can be substituted in function F using an algebraic division procedure (i.e. weak division), then the area value as defined in equation 5.2 is always greater than zero. The important consequence of this is that if a function G can be substituted into a function F, it is guaranteed that this substitution will result in a decrease in the number of literals of the network in the SOP form.

Substitution is also used to minimize the network power consumption. However, substitution does not always guarantee a reduction in the power cost of the network. This means that an operation which decreases the number of literals in the network can potentially increase the power cost of the network in the SOP form.

Example 5.4:

Assume $F_1 = a\,b\,c$ and $F_2 = a\,b$

also $sp(a) = sp(b) = sp(c) = 0.9,\ sp(F_2) = 0.5$

Note that $sp(F_2) \neq sp(a)*sp(b)$. The cause for this is spatial dependence between inputs a and b[4].

F_1 can be expressed in terms of F_2 as follows:

$$F_1 = F_2\,c$$

After the substitution, only the input plane for F_1 is changed in the network, and the power at the input of F_1 is increased from 0.54 to 0.68 even though literal count is decreased.

Once it has been determined that a node n_j can be substituted into a node n_i, the power cost function of n_i is computed as given in equation 5.1 for both SOP implementations of n_i before and after substituting n_j. The substitution is performed if the power cost of n_i is reduced after substitution.

5.5 Selective Collapse

Selective collapse (eliminate) is the process of selectively eliminating nodes in a network in order to reduce the area cost of the network. Selective collapse is performed on an initial Boolean network to provide a better starting point for the extraction procedures. This is done since the initial network might have factors identified which are not good candidates for extraction. During area optimization, two value functions are used to decide if a node should be eliminated by collapsing it into its fanout nodes. One is the sum-of-products value of the node. This value is the reduction in the number of literals in the sum-of-products form of the network if the node is collapsed into all its fanout nodes. The area value and power value for a node n_i are given by equations 5.2 and 5.3.

The second value function is the factored form value. This value gives the reduction in the number of literals in the factored form of the network if the node is collapsed into its fanout nodes. The area value in the factored form of a node is defined as follows [6]:

$$\left(L(n_i) \cdot \sum_{n_k \in fanouts(n_i)} FL(n_i, n_k)\right) - \left(\sum_{n_k \in fanouts(n_i)} FL(n_i, n_k)\right) - L(n_i) \qquad (5.6)$$

where $L(n_i)$ is the number of literals in the factored form of the node n_i. The first term in this equation accounts for the duplication of the literals in the factored expression of n_i once it is collapsed into its fanouts. The second term is the decrease in the number of literals by removing the literals corresponding to n_i from its fanout nodes. The last term account for the removal of n_i from the network where $L(n_i)$ literals are removed from the network.

The factored form power value for selective elimination is defined as follows:

$$\left(\sum_{n_j \in fanins(n_i)} sw(n_j) \cdot FL(n_j, n_i)\right) \cdot \left(\sum_{n_k \in fanouts(n_i)} FL(n_i, n_k) - 1\right)$$
$$- \left(sw(n_i) \cdot \sum_{n_k \in fanouts(n_i)} FL(n_i, n_k)\right) \quad (5.7)$$

The first term in this equation gives an estimate of the power added to the network by duplicating the node function. The second term accounts for the power at the output of the node which is removed from the network.

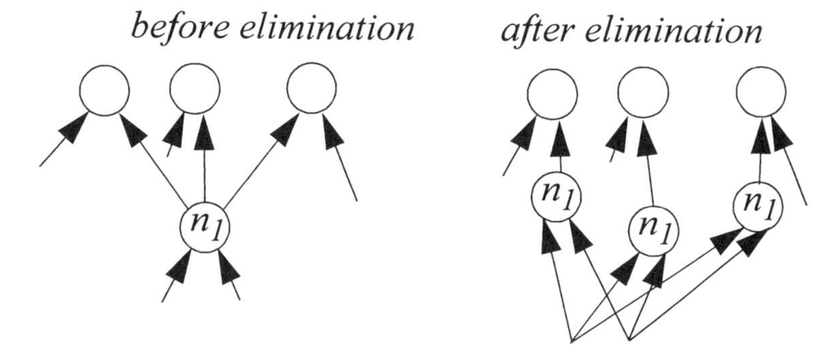

Figure 5.6 Node Elimination in the network

5.6 Experimental Results

The procedures outlined in this chapter have been implemented, and an optimization script called *script.power_alg* has been created.

Each circuit in the benchmark set was first optimized for area using *script.algebraic* [1], and then mapped for minimum power using an industrial library and the power driven technology mapper presented in [8]. The same circuits were then optimized for minimum power using *script.power_alg*, and then mapped using the same library and technology mapper. It is difficult to accurately estimate the switching activity under a real delay model for circuits before technology mapping. Therefore, a zero delay model was adopted for computing the node switching activities. Table 5.1 shows the results when the circuits are optimized using *script.algebraic*. Columns 1 and 2 give the power and area of the circuit after mapping. Column 3 gives the average switching activity of internal nodes in the network after technology mapping. Power is measured using the library loads, and area is measured as the sum of the gate areas. Columns 4, 5, and 6 show the power, area and average switching activity for the nodes in the network before the technology mapping is started. Here, power is measured assuming an SOP implementation and area is given as the number of literals in the network. The results in Table 5.2 are also normalized with respect to the results in Table 5.1. As the results show, power after mapping has been reduced by 21%, at a cost of increasing the area by 7%. The average switching activity has also been reduced by 24%. As shown by the results, the extraction procedure has been very effective in introducing new nodes in the network with small switching activities. There is also a 22% decrease in the total sum of the power for all circuits being optimized, at the expense of increasing the total area by 10%.

Table 5.3 and 5.4 show the same results starting with multi-level examples. Results show an average 10% improvement in power, at the cost of increasing the area by 5%. Average switching activity in the network is also reduced by 12%. The total sum of power over these circuits is also reduced by 6%, at the expense of increasing the area by 6%.

The run-time for *script.power_alg* is much more than that of *script.algebraic*. This is due to the fact that at start up and after each extract (but not during candidate kernel evaluation), global OBDDs were built for computing the switching activities in the network. More efficient techniques have been proposed for computing the network switching activities [4]. Using these efficient methods, the run time of *script.power_alg* can be significantly improved.

ex	Post-Map power	Post-Map area	Post-Map avg sw	Pre-Map power	Pre-Map area	Pre-Map avg sw
apex2	0.88	48.61	0.13	108.85	379	0.223
apex4	4.39	414.08	0.06	496.51	2725	0.118
b12	0.19	11.00	0.14	22.55	93	0.212
clip	0.41	17.15	0.24	50.12	141	0.305
cps	1.80	174.48	0.08	207.71	1187	0.143
duke2	0.86	60.39	0.12	87.88	418	0.159
ex4	1.38	67.35	0.16	176.91	556	0.212
inc	0.46	18.84	0.18	54.66	148	0.300
misex2	0.23	15.74	0.10	23.21	109	0.105
misex3c	0.97	85.58	0.08	111.08	601	0.151
misex3	1.88	118.81	0.11	220.30	844	0.205
pdc	1.47	81.14	0.15	150.19	570	0.177
rd53	0.28	7.00	0.38	33.05	62	0.460
rd73	0.21	19.14	0.10	29.88	143	0.212
rd84	0.43	24.03	0.15	53.49	178	0.275
sao2	0.47	21.75	0.16	60.63	182	0.256
spla	1.39	86.96	0.13	154.14	616	0.172
squar5	0.11	8.96	0.12	13.04	70	0.199

Table 5.1 script.algebraic for two-level MCNC circuits

ex	Post-Map 1	Post-Map 2	Post-Map 3	Pre-Map 4	Pre-Map 5	Pre-Map 6
apex2	0.75	1.18	0.65	0.68	1.08	0.50
apex4	0.74	1.06	0.81	0.72	1.09	0.62
b12	0.96	1.14	0.68	0.92	1.10	0.74
clip	0.79	0.90	0.81	0.82	0.94	0.80
cps	0.85	1.16	0.76	0.79	1.14	0.61
duke2	0.76	1.11	0.75	0.71	1.06	0.52
ex4	0.94	1.07	0.84	0.91	1.05	0.82
inc	0.70	0.92	0.87	0.66	0.95	0.72
misex2	0.94	1.01	0.94	0.95	1.02	0.84
misex3	0.77	1.14	0.65	0.75	1.15	0.55
misex3c	0.78	1.07	0.70	0.82	1.10	0.63
pdc	0.80	1.08	0.67	0.90	1.10	0.73
rd53	0.65	0.81	0.89	0.59	0.69	0.79
rd73	0.78	1.08	0.65	0.72	1.15	0.53
rd84	0.76	1.19	0.63	0.78	1.28	0.56
sao2	0.67	1.19	0.52	0.62	1.03	0.44
spla	0.73	1.16	0.61	0.73	1.11	0.55
squar5	0.80	0.95	0.84	0.84	1.00	0.71
Avg	0.79	1.07	0.74	0.77	1.06	0.65

Table 5.2 script.power_alg for two-level MCNC circuits

ex	Post-Map			Pre-Map		
	power	area	avg sw	power	area	avg sw
C1355	1.80	69.69	0.279	214.14	556	0.290
C1908	1.50	84.49	0.196	170.22	565	0.238
C432	0.62	36.58	0.140	75.06	250	0.178
C880	1.23	57.37	0.199	152.79	445	0.256
alu2	0.88	64.46	0.108	105.00	467	0.208
alu4	1.28	130.25	0.085	102.94	864	0.108
apex6	2.19	113.27	0.167	272.27	827	0.300
b9	0.32	17.40	0.165	38.66	131	0.251
dalu	2.43	158.16	0.147	288.11	1239	0.193
des	8.65	568.79	0.126	1068.61	3765	0.280
f51m	0.35	18.47	0.177	43.28	140	0.298
frg1	0.40	17.88	0.209	46.73	146	0.277
k2	1.39	174.22	0.057	119.84	1069	0.058
rot	1.62	99.89	0.151	204.93	773	0.235
t481	0.85	122.35	0.067	70.93	915	0.052
ttt2	0.47	26.11	0.135	57.41	214	0.188
x2	0.15	6.93	0.184	14.98	54	0.290
9symml	0.61	34.13	0.128	79.68	266	0.262

Table 5.3 script.algebraic for multi-level MCNC circuits

ex	Post-Map			Pre-Map		
	power	area	avg sw	power	area	avg sw
C1355	0.91	0.98	0.87	0.97	1.00	0.95
C1908	0.94	0.90	1.00	0.94	1.01	0.89
C432	0.92	0.90	1.10	0.96	1.01	1.03
C880	0.92	1.00	0.94	0.91	0.99	0.88
alu2	0.85	1.01	0.78	0.87	1.05	0.78
alu4	0.65	1.07	0.57	0.87	1.10	0.68
apex6	0.98	1.05	0.95	0.93	1.06	0.89
b9	0.99	0.99	1.03	0.99	1.08	0.90
dalu	1.07	1.14	0.90	1.12	1.12	0.91
des	0.99	1.02	1.12	0.94	1.05	0.88
f51m	0.89	1.11	0.74	0.96	1.23	0.72
frg1	0.85	1.05	0.72	0.87	1.10	0.72
k2	0.94	1.24	0.89	0.71	1.28	0.57
rot	0.99	1.00	1.02	0.93	1.01	0.91
t481	0.78	1.07	0.68	0.98	0.99	0.82
ttt2	0.85	1.15	0.78	0.86	1.11	0.79
x2	0.89	0.97	1.01	0.88	0.94	0.89
9symml	0.83	1.16	0.75	0.77	1.14	0.61
Avg	0.90	1.05	0.88	0.91	1.07	0.82

Table 5.4 script.power_alg for multi-level MCNC circuits

References

[1] "SIS: A system for sequential circuit synthesis," Report M92/41, UC Berkeley, 1992.

[2] R. Brayton, C. McMullen. "The decomposition and factorization of boolean expressions." In proceedings of the *International Symposium on Circuits and Systems*, pages 49-54, 1982.

[3] R. Brayton, R. Rudell, A. Sangiovanni-Vincentelli, A. Wang. "Multi-level Logic Optimization and the Rectangular Covering Problem" In proceedings of the *IEEE International Conference on Computer-Aided Design*, 1987.

[4] R. Marculescu, D. Marculescu, M. Pedram. "Logic level power estimation considering spatio-temporal correlations." In proceedings of the *IEEE International Conference on Computer Aided Design*, pages 294-299, 1994.

[5] K. Roy, S. C.Prasad. "Circuit activity based logic synthesis for low power reliable operations." In *IEEE Transactions on VLSI Systems*. 1(4):503-513, December 1993.

[6] R. Rudell. "Logic Synthesis for VLSI Design." Ph.D. thesis, University of California, Berkeley, 1989.

[7] A. A. Shen, A. Ghosh, S. Devadas, and K. Keutzer. "On average power dissipation and random pattern testability of CMOS combinational logic networks," In proceedings of the *IEEE International Conference on Computer Aided Design*, November 1992.

[8] C-Y. Tsui, M. Pedram and A. M. Despain. "Power efficient technology decomposition and mapping under an extended power consumption model." In *IEEE Transactions on Computer Aided Design*, Vol.~13, No.~9 (1994), pages 1110--1122.

[9] A. Wang. "Algorithms for Multi-Level Logic Optimizations." Ph.D. Thesis, UC Berkeley, 1989.

CHAPTER 6

Logic Minimization for Low Power

A multi-level logic circuit can be modeled as a Boolean network where each node in this network has a Boolean function represented by a SOP form equation and each edge represents a signal connection. It is possible to obtain a multi-level circuit from either the HDL representation of a design or by applying logic restructuring operations on a two-level circuit representation.

Two level function minimization techniques as described in chapter 3, can be used to minimize the local function of each node in a Boolean network. In a multi-level network, don't care conditions are used to provide more flexibility in minimizing the local function of each node. Current techniques for computing and using these don't care conditions allow for correct functional operation of the network by guaranteeing that as each function in the network is being optimized using its don't care conditions, other functions in the network are changed only within their don't care sets. Although this approach is consistent with the notion of area optimization, it cannot be directly applied to power optimization. As stated before, as local function a node is being changed during local node minimization, it is possible for the global function of other nodes in the network to be changed within their don't care conditions. Since switching activity values in a network depend on global functions of nodes in the network, local node minimizations have a global impact on the network power consumption. This observation clearly shows the need for new methods for analyzing the effect of using don't care sets on the switching activity of nodes in the network.

This chapter provides new approaches for computing observability conditions in a network which take into account the power consumption of the network. These techniques are then used to propose new techniques for computing, and using don't care sets, which result in power optimal networks.

The problem of don't care computations and Boolean function minimization has been addressed by many researchers in the past. ESPRESSO [3] presents a heuristic approach where novel techniques are used to efficiently produce good area solutions, while ESPRESSO EXACT [14] presents an exact method for solving the minimum area solution. In [6] a method is presented for computing the complete set of observability conditions for each node in the network where the observability don't care (*ODC*) at each node is computed as a function of the *ODC* for its fanout edges.

The problem of power optimization during logic synthesis has been addressed in a number of publications. The don't care set computed for area optimization [15] is used in [16] to optimize the local function of nodes for power. This work does not, however, take into account the effect of changes in the function of internal nodes on the power consumption of other nodes in the network. The idea of power relevant don't cares was first introduced in [9], where an efficient technique for computing power relevant don't care conditions was presented. The computed power relevant don't care guarantees that any changes in the local function of a node does not result in increasing the switching activity of other nodes in the network beyond their value, *when these nodes were optimized*. The notion of minimal variable supports is used in [9] to optimize the local function of nodes for power.

The analysis on power relevant don't cares [9] is used in [10] to compute a resynthesis potential for nodes in a technology mapped network. This resynthesis potential represents the estimated effect of a change in the local function of a node on the power consumption of its transitive fanout nodes. The method presented in [10] also takes into account changes in power consumption due to variations in hazardous transitions in the network after an internal node is resynthesized.

This first part of this chapter presents a method for computing the power relevant observability and satisfiability don't cares for nodes in a Boolean network. The power relevant observability don't care presented here consists of the power relevant don't cares presented in [9]. In addition, an extension to this theory is presented that makes possible computation of Monotone Power Relevant Observability Don't Cares. Using this don't care set, it is guaranteed that local node optimization does not increase the switching activity of other nodes, *beyond their current value*. This results in a monotone decrease in the switching activity of the network as nodes are optimized. The power relevant satisfiability don't care conditions included in the don't care set, are intended to maximize the probability of substituting nodes with low switching activity into the node being optimized. The approach to compute the satisfiability don't care [15] is modified in this chapter to include low switching activity nodes (from which a subset of nodes can be selected for possible substitution into the function of node being optimized).

The second part of the chapter uses the notion of minimal variable supports to find a minimal power implementation of the node function. Using this approach, and also using the power relevant satisfiability don't cares described above, inputs with high switching activity may be replaced with new fanins with lower switching activities when permitted by the power relevant don't care conditions derived from the network functionality and structure.

The balance of this chapter is organized as follows. Section 6.1 presents techniques for computing the set of power relevant don't cares. Power relevant don't cares are then used in section 6.2 to present a method for optimizing the local function of nodes for low power using the concept of minimal variable supports. Experimental results are presented in section 6.3.

6.1 Power Relevant Don't Care Conditions

Logic synthesis algorithms use the flexibility provided by don't care conditions to more effectively manipulate a Boolean network. Current techniques for computing and using these don't care conditions allow for correct functional operation of the network by guaranteeing that as each function in the network is being optimized using its don't care conditions, other functions in the network are changed only within their don't care sets.

Power consumption in a Boolean network is proportional to the switching activity of the nodes in the network. Switching activity of a node is a function of the signal probability of the function and therefore dependent on the global function of the node. Therefore, as a node is being optimized using its don't care set, the switching activity of the node function, as well as the switching activity of other nodes in the network, are also being changed. This clearly shows the need for new methods to analyze the effect of using don't care sets on the switching activity of nodes in the network. This section provides such an analysis. The analysis presented in this section is used to propose new techniques for computing and using power relevant observability don't care sets. Power relevant satisfiability don't cares are the subject of section 6.1.7.

6.1.1 Don't care conditions for area

In logic synthesis, the concept of don't cares is used to represent the available flexibility in implementing Boolean functions. Don't care conditions for a function specify part of the Boolean space where the function can evaluate to one or zero. Three sources of don't cares are external don't cares (*EDC*), observability don't cares (*ODC*) and satisfiability don't cares (*SDC*). It has been shown that if a node is minimized using all three types of don't cares, then all connections to, and inside the node, are irredundant. This means that *EDC*, *ODC* and *SDC* provide a complete set of don't care conditions during optimization [2].

Definition 6.1 *The external don't care set for each output* z_i *of the network is all input combinations that either do not occur or the value of* z_i *for that input combination is not important.*

Definition 6.2 *If* y_i *is the variable at an intermediate node and* f_i *its logic function, then* $y_i = f_i$. *Therefore, we don't care if* $y_i \neq f_i$. *The expression* $SDC = \sum (y_i \oplus f_i)$ *for all nodes in the network is called the Satisfiability Don't Care set (SDC).*

Definition 6.3 *The observability don't care set (ODCs) at each intermediate node* y_0 *of a multi-level network are conditions under which* y_0 *can be either 1 or 0 while the functions generated at each primary output remain unchanged. If* $z = (z_1, \ldots, z_l)$ *gives the set of circuit outputs, then the complete ODC at node* y_0 *is:*

$$ODC_0 = \langle m \in B^n | z_{y_0}(m) \equiv z_{\bar{y}_0}(m) \rangle = \prod_{i=1}^{l} \overline{\frac{\partial z_i}{\partial y_0}} \tag{6.1}$$

6.1.1.1 Computing Don't Care Conditions

The complete don't care set for a node *n* is found by first computing the *ODC* as a function of the primary inputs of the circuit. The external don't care, which is also expressed in terms of the primary inputs, is then added to *ODC*. Image projection techniques [5] are then used to find the *ODC* plus *EDC* of the node in terms of the immediate fanins of the node. Finally a subset of *SDC* for nodes which can be substituted into *n* with high probability is added to this local don't care. In general, computing the *ODC* for a node is the most complex part of this computation. The procedure for computing *ODC* conditions is briefly described in the following.

In [6], a method is described for computing the complete set of observability conditions for each node in the network where the *ODC* at each node is computed as a function of the *ODC* for its fanout edges. In this procedure, the *ODC* of the node with respect to each primary output, is computed separately. A different technique for computing the complete don't care set is presented in [15], which takes advantage of observability relations [4] at the primary outputs of the network. The given algorithm computes the complete *ODC* for each node in a multi-level combinational network. For tree networks, the following equation is used to compute the maximal set of *ODC*s at the output a node *g*.

$$ODC_g = ODC_f + ODC_g^f \tag{6.2}$$

where $ODC_g^f = \overline{\frac{\partial f}{\partial g}}$.

The complete *ODC* cannot, however, be used in synthesis for any real size problem. This is because once the function of a node is minimized using its complete *ODC*, the *ODC* at other nodes in the network will potentially change. Therefore the *ODC* for all nodes have to be recalculated after each optimization step. At the same time, the size of the complete *ODC* can become extremely large. Therefore, subsets of *ODC* have to be used instead of the complete *ODC* of each node. *RESTRICT* was the first *ODC* filter introduced in [8]. This filter removed any cube in the *ODC* of a node y_i which had a literal corresponding to a node in its transitive fanout. Although this filter and a number of other filters made the *ODC*s smaller, *ODC*s still had to be recomputed after each node simplification. Compatible set of permissible functions (*CSPF*), introduced in [13] allows for simultaneous optimization of all nodes in a network. In [15] the concept of *CSPFs*, which is only defined for NOR gates, is extended to complex nodes of a general multi-level network and is called Compatible *ODC*s (*CODCs*). *CODCs* are used to simultaneously minimize the function of each node in the network. Even though by using *CODCs* some of the information contained in each of the full *ODCs* is lost, the time complexity of using the full *ODC* makes it impossible to use them for any practical size problem. The following equation presented in [15] is used for calculating the maximal set of *CODC* for the fanins of a node *g* in a tree network.

$$CODC_g = CODC_g^f + CODC_f \tag{6.3}$$

Two methods are also presented for computing $CODC^f_g$. The first method produces the maximal set for $CODC^f_g$ but is shown to be too expensive. The second method shown by the following equation is a more efficient technique for computing a valid subset of the maximal set of $CODC^f_g$.

$$CODC^f_g = \left(\frac{\partial f}{\partial y_1} + C_{y_1}\right) \ldots\ldots \left(\frac{\partial f}{\partial y_{k-1}} + C_{y_{k-1}}\right)\overline{\frac{\partial f}{\partial y_k}} \tag{6.4}$$

In this equation g is assumed to be the *kth* input y_k to function f and also the variable ordering $y_1 > \ldots . > y_r$, is assumed for the fanins of function f. Also, $C_x(f) = f_x \cdot f_{\bar{x}}$ is the largest function independent of variable x which is contained in function f.

These results are then generalized to multiple fanout nodes by finding the intersection of the compatible *ODCs* for each fanout edge. This operation is shown by the following equation:

$$CODC_g = \prod_{f_i \in fanouts} CODC_g(f_i) \tag{6.5}$$

where $CODC_g(f_i)$ gives the compatible *ODC* computed using equation 6.3 for fanout f_i of node g.

This method for computing the compatible don't care set requires that each node be optimized only after all its fanout nodes have been optimized. The procedure shown in figure 6.1 is used to optimize the function of nodes in the network using the computed don't care set.

Function *computeCODC* uses equation 6.5 to compute the compatible observability don't care in terms of the primary inputs. Function *computeLocalDC* will then find the don't care in terms of the immediate fanins of the function using image projection techniques. The function of the node is then optimized using the local *CODC* and a subset of the SDC for nodes which, with high probability, may be substituted into the node.

6.1.2 Observability don't care conditions for power

The compatible don't care computed in section 6.1.1 is freely used while minimizing the function of nodes in a Boolean network, guaranteeing that the global function of circuit outputs will only change within their specified external don't care set. The change in the global function of transitive fanouts of node *n* as *n* is optimized, is not of concern when area is being minimized since the change in the function of each node will be within the observability don't care calculated for that node. However, as mentioned before, this observation does not hold for power minimization. For example, if by modifying the function of an internal node, the signal probability of a fanout node is changed from 0.1 to 0.2, we can expect 78% increase in the power consumption of the fanout node. The following example shows how optimizing the function of an internal node for power may result in an increase in the total power consumption of a network.

Example 6.1:

Consider a function *f* with a load of 5 and function *g* with load 1 where:

$$f = g + a\,;$$

$$g = \bar{a}\,.\,b;$$

Also: $dc_f = \bar{a}.b$, *also* $sp(a)=sp(b)=0.5$

Then: $DC_g = dc_f + a = a + b$

Before optimization:

$$f = a + \bar{a}.b \Rightarrow sp(f) = 3/4 \Rightarrow sw(f) = 3/8 \Rightarrow Power(f) = 15/8$$

$$g = \bar{a}.b \Rightarrow sp(g) = 1/4 \Rightarrow sw(f) = 3/8 \Rightarrow Power(g) = 3/8$$

After optimization, function *g* is set to *0*:

$$f = a \Rightarrow sp(f) = 2/4 \Rightarrow sw(f) = 4/8 \Rightarrow Power(f) = 20/8$$

$$g = 0 \Rightarrow sp(g) = 0 \Rightarrow sw(f) = 0 \Rightarrow Power(g) = 0$$

```
1:  function full_simplify(G)
2:    G(V, E) is a Boolean network;
3:    begin
4:        for node ∈ G in reverse dfs order do
5:            CODC  = computeCODC(node)
6:            CODC^l = computeLocalDC(node, CODC)
7:            SDC = selectNodeSDC(node)
8:            newf = nodeSimplify(node, CODC^l, SDC)
9:        endfor
10:   end
```

Figure 6.1 Computing compatible *ODCs* in a network

This example shows that after optimizing node g, the power at the output of node f is increased.

6.1.3 Observability don't care analysis for power optimization

This section presents an analysis of the effect of using observability don't cares on the signal probability of other nodes in a tree network. This analysis is then used to derive exact and heuristic methods for computing power relevant observability don't cares.

6.1.3.1 Observability don't care analysis for tree networks

Assume a function f and its fanin g. The relationship between the global function g and its maximal *ODC* set is shown in Figure 6.2. The space of the primary inputs is shown by the area inside the large circle. The area inside the smaller circle represents the on-set points for function g. The single dashed region gives all conditions where changes in function g is unobservable at node f. The combination of the single and cross dashed regions give all the primary input combinations where changes in g is unobservable at the primary outputs. Note that $g \cap ODC_g \neq \emptyset$.

The following lemmas are used to study the effect of changes in the function of node g in the function of node f.

Lemma 6.1 *$\partial f^+/\partial g$ (positive dependency) as defined here, gives all global conditions where both functions f and g evaluate to the same value.*

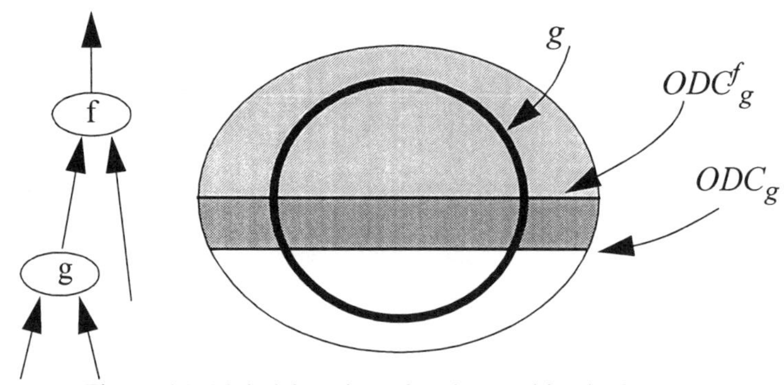

Figure 6.2 Global function of node g and its don't cares

$$\frac{\partial f^+}{\partial g} = f_g \cdot \overline{f_{\bar{g}}} \qquad (6.6)$$

Proof: The following relations exists between f and g where f_g is the cofactor of f with respect to variable g:

f_g: global conditions where $f=1$ if $g = 1$

$f_{\bar{g}}$: global conditions where $f=0$ if $g = 0$

The intersection of these two functions gives all conditions where f changes from 0 to 1 when g changes from 0 to 1. In other words:

$$\{ \partial f^+/\partial g \subseteq B^n \mid \forall v \in B^n, f(v) = g(v)\}.$$

■

Lemma 6.2 *$\partial f^-/\partial g$ (negative dependency) as defined here gives all global conditions where functions f and g evaluate to opposite values.*

$$\frac{\partial f^-}{\partial g} = \overline{f_g} \cdot f_{\bar{g}} \qquad (6.7)$$

Proof: The following relations exists between f and g:

$f_{\bar{g}}$: global conditions where f =1 if g = 0

f_g: global conditions where f =0 if g = 1

The intersection of these two functions gives all conditions where f changes from 1 to 0 when g changes from 0 to 1. In other words:

$$\{ \partial f^-/\partial g \subseteq B^n \mid \forall v \in B^n, f(v) = \bar{g}(v)\}.$$

■

Note that $(\partial f/\partial g = \partial f^+/\partial g + \partial f^-/\partial g)$ which is the equation for the difference of f with respect to g. Note also that $(\partial f^+/\partial g \cap \partial f^-/\partial g) = \emptyset$.

Figure 6.3 shows how the global space of the primary inputs is partitioned with respect to global functions of f and g. The region specified by ODC^f_g specifies all points of the global space where changes in g will not affect the global function of f. Region $\partial f^+/\partial g$ specifies all points in the global space which if included in the on-set of g will also be included in the on-set of f. This means that by including these points in g, the number of minterms in f will also increase (hence the up arrow). Region $\partial f^-/\partial g$ specifies all the points in the global space which if included in the on-set of g

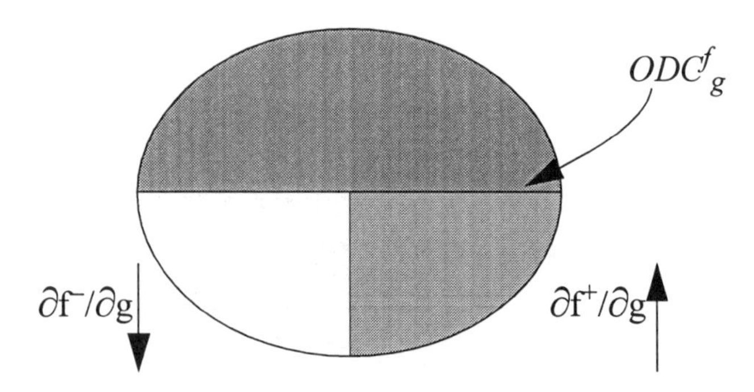

Figure 6.3 Partition of the space of primary inputs for function f and g

will be removed from the on-set of *f*. This means that including these points in *g* will reduce the number of points in the onset of *f* (hence the down arrow).

6.1.3.2 Observability Don't Care Regions

Figure 6.4 shows the relationship between the global functions of $\partial f^- / \partial g$, $\partial f^+ / \partial g$ and *g*. This figure is obtained by overlapping figures 6.2 and 6.3. Even though no assumption is made about the global functions of nodes *f* and *g*, figure 6.4 shows all possible combinations of the regions shown in figures 6.2 and 6.3 assuming a tree network.

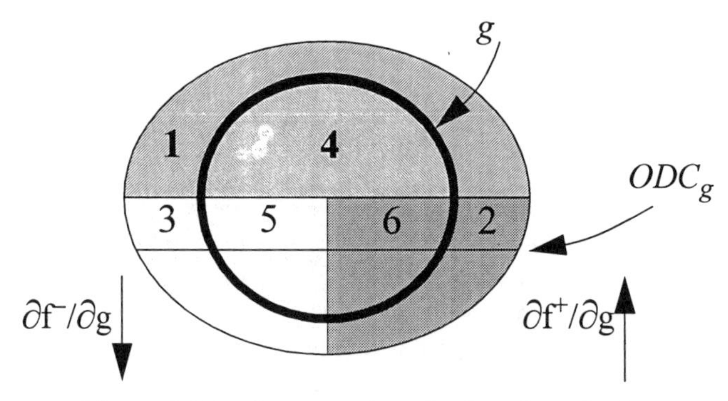

Figure 6.4 Don't care regions for functions *f* and *g*

The global don't care conditions for node g (region above line ODC_g in figure 6.4) is partitioned into six regions. Each of these regions specifies a don't care region of g with respect to f as defined in the following.

Definition 6.4 *Given a node g and its fanout node f, the don't care regions of g with respect to f are denoted as* $R_{g,f}(\alpha, \beta)$. *This don't care region specifies all global conditions where g evaluates to* α $(\alpha=\{0,1\})$. *The second entry,* $\beta=\{0,1,-\}$ *specifies whether for points in this region f evaluates to the same value as g* $(\beta=0)$, *the opposite value of g* $(\beta=1)$ *or whether f is independent of g* $(\beta= -)$.

For example region 3 in figure 6.4 is denoted as $R_{g,f}(0,1)$ and region 4 is given as $R_{g,f}(1,-)$. The definition for don't care regions is extended in the following to represent the relationship between a node g and its transitive fanouts nodes.

Definition 6.5 *Given a node g and its fanout nodes* $F=\{f_1,...f_k\}$, *the don't care regions of g with respect to F are denoted as* $R_{g,F}(\alpha, \beta)$. *This don't care region specifies all global conditions where g evaluates to* α $(\alpha=\{0,1\})$. *The second entry is a k bit vector where each bit takes values from the set* $\beta=\{0,1,-\}$. *Bit i of this vector specifies whether for points in this region* f_i *evaluates to the same value as g* $(\beta_i=0)$, *an opposite value than g* $(\beta_i=1)$ *or whether* f_i *is independent of g* $(\beta_i=-)$.

The following lemmas give properties of don't care regions which will be used to study the effect of changing the global function of node g on the global function of its fanout node f.

Lemma 6.3 *Using the minterms in* $R_{g,f}(1,-)$ *and* $R_{g,f}(0,-)$ *while optimizing node g will not affect the global function of node f.*

Proof: The proof follows immediately from the definition for don't care regions.

■

Lemma 6.4 *Using a minterm* v_i *in don't care regions* $R_{g,f}(1,1)$ *and* $R_{g,f}(0,0)$ *while optimizing node g will result in bringing this minterm from the off-set to the on-set of function f.*

Proof: For all minterms in region $R_{g,f}(0,0)$ ($R_{g,f}(1,1)$) function g evaluates to $0(1)$. This means that using a minterm in this region while optimizing function g is equivalent to including this minterm in the on-set (off-set) of node g. Also for all minterms in this region, f and g evaluate to the same(opposite) values. This means that for all minterms in this region function f evaluates to 0. Therefore including a minterm in this region in the on-set (off-set) of g will also include this minterm in the on-set of f.

■

Lemma 6.5 *Using a minterm* v_i *in don't care regions* $R_{g,f}(0,1)$ *and* $R_{g,f}(1,0)$ *while*

region	function g	function f
$R_{g,f}(0,-)$	including v in g	f is not changed
$R_{g,f}(0,0)$	including v in g	v is included in f
$R_{g,f}(0,1)$	including v in g	v is excluded from f
$R_{g,f}(1,-)$	excluding v from g	f is not changed
$R_{g,f}(1,1)$	excluding v from g	v is included in f
$R_{g,f}(1,0)$	excluding v from g	v is excluded from f

Table 6.1 Effect of changes in global function of g on global function of f

optimizing node g will result in bringing this minterm from the on-set to the off-set of function f.

Proof: proof is similar to the previous lemma.

■

Table 6.1 summarizes the results of these lemmas. In this table, v is a minterm in the space of the primary inputs of the network.

For each region in ODC_g, the change in the function of f as minterm v_i in region i is included in, or excluded from the on-set of g is well defined. This means that while optimizing node g, the effect of changes in global function of node f is exactly known using the information on the don't care regions for node g. Therefore, the effect of changing the function of g on the signal probability and, therefore, the switching activity of node f can exactly be measured. In the next section we discuss how don't care regions are used to optimize the function of a node, while considering the global effects of this change.

6.1.3.3 Using Don't Care Regions in Node Optimization

During area optimization, the local don't care set of a node is used to optimize the local function of the node for area. During network optimization for power, the don't care information for a node will have to be used to minimize the combination of the power contribution of the node to the network power, as well as the switching activity in the transitive fanout nodes.

In a tree network, most nodes have more than one node in their transitive fanouts. This means that while optimizing a node, it is necessary to consider the effect of changes in the function of this node on all nodes in its transitive fanouts. The following lemma gives the number of don't care regions for a node in a tree network.

Lemma 6.6 *For a node g in a single output tree network with k nodes in its transitive fanout, the number of don't care regions is given by:*

$$2 \cdot \left(3^k - \sum_{j=0}^{k-1} j \cdot (3^{k-j} - 1) \right) \tag{6.8}$$

Proof: The maximum number of don't care regions for node g with respect to its k fanout nodes is given as 2.3^k. This is the possible number of combinations for specifying $R_{g,F}(\alpha, \beta)$ as given in definition 6.5. For example if $k=2$ then the following enumerates all possible don't care regions for $g=0$:

$R_{g,F}(0,00)$	$R_{g,F}(0,01)$	$R_{g,F}(0,0\text{-})$
$R_{g,F}(0,10)$	$R_{g,F}(0,11)$	$R_{g,F}(0,1\text{-})$
$R_{g,F}(0,\text{-}0)$	$R_{g,F}(0,\text{-}1)$	$R_{g,F}(0,\text{--})$

In a tree network, however because of the relationships between observability conditions, some don't care regions will be empty. For example, assume $F=(f,h)$ where f is fanout of g and h is fanout of f, Then $R_{g,F}(0,\text{-}0)$ will always be empty. This is because this don't care region refers to points where changes in the value of g do not affect the function of f but it does affect the function of h. This is not possible in a tree network.

Assume the k fanouts of node g specified in F are ordered such that each fanout is listed before all its fanouts. Then don't care regions for $\beta = (\beta_1, \ldots, \beta_{i-1}, \text{"-"}, \beta_{i+1}, \ldots, \beta_k)$ are not empty only if none of the values in $\beta_1, \ldots, \beta_{i-1}$ are "-" and all values in $(\beta_{i+1}, \ldots, \beta_k)$ are all "-". Computing the number of all such conditions and then multiplying it by 2 (for g being 0 or 1) proves the claim of this lemma.

■

By using don't care regions for node g and nodes in its transitive fanout, the effect of changes in the function of node g on the signal probability of these fanout nodes can be analyzed.

There are two drawbacks in using the don't care regions to minimize the switching activity of a node and its transitive fanout nodes. The first drawback is that in optimizing node g, information about all don't care regions corresponding to its transitive fanout nodes is needed and since the number of don't care regions grows exponentially with the number of fanouts (lemma 6.6), this analysis very quickly becomes

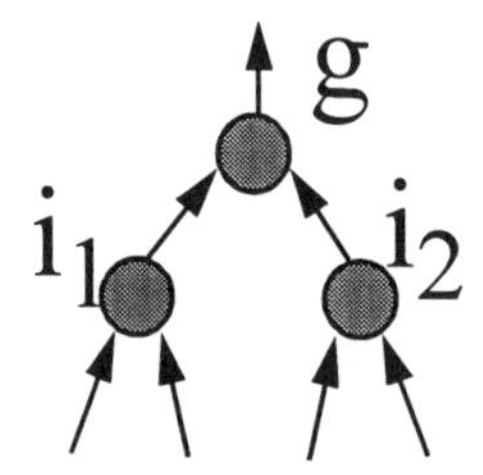

Figure 6.5 Contradictory decisions on probability of g while optimizing i_1 and i_2

intractable. A second problem is that contradictory decisions as to increase or decrease the signal probability of a node g can be made while optimizing nodes i_1 and i_2 in its transitive fanin (figure 6.5). This means that even if all the optimization problems are solved, it is still possible to obtain no improvement because of increasing the signal probability of a node at one step and decreasing it during another step of the procedure.

The complexity of power optimization procedure is reduced if a decision is made to only increase, or decrease, the signal probability of a node function after it has been optimized. An added advantage of this approach is that conflicting decisions will not be made regarding the new signal probability of this node since nodes in its transitive fanin are being optimized.

The following theorems are used to reduce the complexity of the power optimization procedure.

Theorem 6.1 *Given a node g and its fanout node f, $R_{g,f}(0,1)$ and $R_{g,f}(1,0)$ are the empty sets and $R_{g,f}(0,0)$ and $R_{g,f}(1,1)$ are maximal if ODC_g is computed as:*

$$ODC_g = ODC_g^f + \overline{f} \cdot ODC_f \tag{6.9}$$

Proof: The following equations are derived from definition:

$$R_{g,f}(0,0) = \bar{g} \cdot f_g \cdot \overline{f_{\bar{g}}} \cdot ODC_g$$

$$R_{g,f}(0,1) = \bar{g} \cdot f_{\bar{g}} \cdot \overline{f_g} \cdot ODC_g$$

$$R_{g,f}(1,0) = g \cdot f_g \cdot \overline{f_{\bar{g}}} \cdot ODC_g$$

$$R_{g,f}(1,1) = g \cdot f_{\bar{g}} \cdot \overline{f_g} \cdot ODC_g$$

Or equivalently (by substituting $ODC_g = ODC^f_g + ODC_f$):

$$R_{g,f}(0,0) = \bar{g} \cdot f_g \cdot \overline{f_{\bar{g}}} \cdot ODC_f$$

$$R_{g,f}(0,1) = \bar{g} \cdot f_{\bar{g}} \cdot \overline{f_g} \cdot ODC_f$$

$$R_{g,f}(1,0) = g \cdot f_g \cdot \overline{f_{\bar{g}}} \cdot ODC_f$$

$$R_{g,f}(1,1) = g \cdot f_{\bar{g}} \cdot \overline{f_g} \cdot ODC_f$$

which give maximal such conditions.

Now if equation 6.9 is used instead and considering that $f = g \cdot f_g + \bar{g} \cdot f_{\bar{g}}$:

$$R_{g,f}(0,0) = \bar{g} \cdot f_g \cdot \overline{f_{\bar{g}}} \cdot ODC_f$$

$$R_{g,f}(0,1) = 0$$

$$R_{g,f}(1,0) = 0$$

$$R_{g,f}(1,1) = g \cdot f_{\bar{g}} \cdot \overline{f_g} \cdot ODC_f$$

which shows that regions $R_{g,f}(0,1)$ and $R_{g,f}(1,0)$ are the empty sets and $R_{g,f}(0,0)$ and $R_{g,f}(1,1)$ are maximal. This proves the claim of this lemma.

■

Theorem 6.2 *Given a node g and its fanout node f, $R_{g,f}(1,1)$ and $R_{g,f}(0,0)$ are the empty sets and $R_{g,f}(0,1)$ and $R_{g,f}(1,0)$ are maximal if ODC_g is computed as:*

$$ODC_g = ODC^f_g + f \cdot ODC_f \tag{6.10}$$

Proof: Proof is similar to proof for theorem 6.1.

■

Theorems 6.1 and 6.2 are used as follows. Assume that after optimizing a node *f*, we decide that as other nodes in the network are optimized we want the signal probability of this node to remain the same, or only increase beyond its current value. This, for example, is desirable when the signal probability of *f* after it is optimized, is more than 0.5. This condition will disallow any increase in the switching activity of the node beyond its current value. Since a decision is made to only allow an increase in the signal probability of function *f*, only regions $R_{g,f}(0,-)$, $R_{g,f}(1,-)$, $R_{g,f}(0,0)$ and $R_{g,f}(1,1)$ need to be used while optimizing a node *g* in the fanin of *f*. Using theorem 6.1, ODC_g is computed such that regions $R_{g,f}(0,1)$ and $R_{g,f}(1,0)$ are empty sets and regions $R_{g,f}(0,0)$ and $R_{g,f}(1,1)$ are maximal. This means that while optimizing node *g*, any use of ODC_g will only result in an increase in the signal probability of node *f*, and hence, the function of node *g* can be optimized without any concern about adverse effects on switching activity of node *f*. The same approach is used to apply theorem 6.2 when disallowing an increase in the signal probability of node *f* (this is desirable if the signal probability of node is less than *0.5* after it is optimized).

This discussion motivates the definition for *Propagated Power Relevant Observability Don't Care* conditions for a node *f*.

Definition 6.6 *Given a node f and its fanin node g, $PPODC_f$, the Propagated Power Relevant Observability Don't Care for node f is defined as a subset of the observability don't care conditions for f that is used to compute the observability don't care conditions for node g while guaranteeing that any changes in the function of g does not increase the switching activity of node f. $PPODC_f$ is defined as follows:*

$$PPODC_f = \begin{cases} \overline{f} \cdot ODC_f & sp(f) > 0.5 \\ f \cdot ODC_f & sp(f) \leq 0.5 \end{cases} \tag{6.11}$$

6.1.4 Power relevant observability don't cares for power optimization

The goal of don't care computation approach outlined in the previous section is to allow the optimization procedure to concentrate on local changes without concern about global effects on power. This is equivalent to providing maximum flexibility in optimizing a node, while guaranteeing that local optimizations do not degrade total power of the network. The analysis presented in the previous section is used here to propose a method for computing the set of observability don't care conditions which guarantee that any changes in the function of an internal node using this don't care set does not increase the switching activity of its immediate fanout node. The impact of using this don't care on the switching activity of nodes other than the immediate fanout of the node that is being optimized is then discussed. In the next section, the technique presented here is extended to guarantee that changing the function of an internal node does not increase the switching activity of any other node in the network. The following definition is provided for the observability don't care set computed in this section:

Definition 6.7 *Given a node g and its immediate fanout node f (figure 6.6), $PODC_g$, the Power Relevant Observability Don't Care for node g is defined as the observability don't care conditions for g that guarantees any changes in the function of g does not increase the switching activity of its immediate fanout node f.*

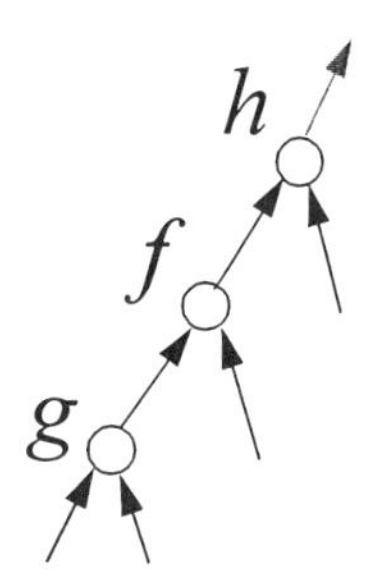

Figure 6.6 Computing Power Relevant Observability Don't Care for a node g

The following equation is used to compute the power relevant observability don't care for an internal node of Boolean network.

$$PODC_g = ODC_g^f + PPODC_f \tag{6.12}$$

In this equation, ODC^f_g gives all conditions that make changes in g unobservable in the function of f while $PPODC_f$ gives the maximal conditions that guarantee that any changes in the of function g will not increase the switching activity of node f (theorems 6.1 and 6.2).

This equation provides a recursive procedure for computing the *PODC* of all nodes in a tree network where each node is optimized after all its fanout nodes. The procedure starts at the primary outputs of the network where *PODC* is equal to the external don't care defined by the user. After a node f is optimized using $PODC_f$, $PPODC_f$ (using equation 6.11) and *f_new* (the new function of node f after optimization) are computed. $PPODC_f$ is then stored at node f for future reference. When node g, fanin node of f, is being optimized, $PPODC_f$ that was stored at node f is used to compute $PODC_g$. The same procedure is recursively applied until all the nodes are optimized. Figure 6.7 shows the procedure for computing the $PPODC_f$ once the function of node f is optimized.

Example 6.2: Reconsider example 1.

Assume $\quad dc_f = \bar{a}.b, \;\; also\ sp(a) = sp(b) = 0.5$

Then $\quad PODC_g = PPODC_f + a = f.dc_f + a = 0 + b = a$

After optimization, function g is either set to b or is not changed. Regardless of how g is optimized, the switching activity of f does not increase.

```
1:  function find_propagated_power_odc(new_f, ODC_f)
2:    new_f is the node function after optimization
3:    PODC_f is the Power Relevant observability don't care for f
4:     begin
5:       if ( sp(new_f) > 0.5 ) then
6:            PPODC_f = new_f . PODC_f
7:       else
8:            PPODC_f = new_f . PODC_f
9:       endif
10:    end
```

Figure 6.7 Computing propagated power relevant don't cares

The procedure presented above guarantees that as a node g is optimized, the switching activity of its immediate fanout node f does not increase (figure 6.6). This procedure does not, however, make any assumptions on the effect of changing g on the switching activity of nodes in the transitive fanout of node f (e.g. node h), as explained next.

Let us assume that after node h is optimized, we decide to increase its signal probability when its fanin nodes are being optimized. This means that don't care regions $R_{f,h}(0,-)$, $R_{f,h}(1,-)$, $R_{f,h}(0,1)$ and $R_{f,h}(1,0)$ are used when optimizing node f. After optimizing node f, the global function of node h changes from 0 to 1 when the function of node f changes from 1 to 0 for set of points $V_{h,1}$ in region $R_{f,h}(1,0)$ and from 0 to 1 for the set of points $V_{h,0}$ in region $R_{f,h}(0,1)$. Also assume after optimizing the function of node f, we decide to increase the signal probability of this node when its fanin node is being optimized. This means that regions $R_{g,f}(0,-)$, $R_{g,f}(1,-)$, $R_{g,f}(0,1)$ and $R_{g,f}(1,0)$ are used in optimizing the function of node g. The function of node f changes from 0 to 1 when the function of node g changes from 1 to 0 for set of points $V_{f,1}$ in region $R_{g,f}(1,0)$ and from 0 to 1 for the set of points $V_{f,0}$ in regions $R_{g,f}(0,1)$. Now, if the intersection of $V_{h,1}$ and $(V_{f,0} + V_{f,1})$ is not empty, then changing the function of f from 0 to 1 will result in changing the function of h from 1 to 0 which results in decreasing the signal probability of node h. This analysis shows that it is possible to increase the switching activity of node h while optimizing node g. Note, however, that any changes in the function of g will not reduce the signal probability of h below its value when it was optimized. This is because the changes in function of g will only affect the function of h in the space defined by don't care regions $R_{f,h}(0,1)$ and

$R_{f,h}(1,0)$ and within this subspace, the function of h only evaluates to 1 for points in region specified by $(V_{h,0} + V_{h,1})$, which was only changed from 0 to 1 after node h was optimized.

This analysis shows that after network optimization using the procedure presented in this section, the switching activity of each node in the network is less than, or equal to, its switching activity immediately after it was optimized. This means that while optimizing different nodes in the transitive fanin of a node f, it is possible to increase or decrease the switching activity of f. However, this new value is always no larger than what it was when node f was first optimized. The next section presents a method that guarantees a monotone decrease, or no change in the current switching activity of all nodes in the network as local node functions are being optimized.

6.1.5 Monotonic reduction in global power

As discussed in the previous section, it is desirable to obtain a monotonic decrease in the global power consumption during node optimizations. The following lemma is used in developing a don't care computation technique that achieves this goal.

Definition 6.8 *Given a node g and its immediate fanout node f (figure 6.6), $MPODC_g$, the Monotone Power Relevant Observability Don't Care for node g is defined as the observability don't care conditions for g that guarantee changes in the function of g within this don't care set, do not increase the current switching activity of nodes in the transitive fanout cone of f (excluding node f).*

Lemma 6.7 *Consider nodes f and h (figure 6.6). Assume $MPODC_f$ has been computed and f has been optimized using $MPODC_f$ resulting in new_f. $MODC_f$, as defined here, gives the maximal set of conditions that when used in computing the observability don't care conditions for the fanins of f, guarantees that any changes in the transitive fanin cone of f, does not increase the switching activity of any node in the transitive fanout cone of f.*

$$MODC_f = ODC_f^h + MPODC_f \cdot \overline{(f \oplus new\_f)} \qquad (6.13)$$

Proof: $(f \oplus new\_f)$ gives all conditions where function of node f has been changed. Note that this change is contained within $MPODC_f$. Therefore, it is also guaranteed that this change has resulted in a decrease in the current switching activity of all transitive fanout nodes where this change is observable. This is a direct consequence of the property defined for $MPODC_f$. For power optimization, if we disallow the function of f to change for all points within $(f \oplus new\_f)$, we are guaranteed that the power reduction obtained in the transitive fanout by changing f will be maintained. This means that $MODC_f$ is obtained by removing the set $(f \oplus new\_f)$ from part of $MPODC_f$ which makes changes in f observable at the transitive fanout cone of f. The part of $MPODC_f$ that makes changes in f observable at the transitive fanout cone of f is given by:

$$MPODC_f \cdot \overline{ODC_f^h}$$

which is all points in $MPODC_f$ that are not in $ODC^h{}_f$.

Then, considering that $ODC^h{}_f \subseteq MPODC_f$:

$$MODC_f = MPODC_f - \left\{ MPODC_f \cdot \overline{ODC_f^h} \cdot (f \oplus new\_f) \right\}$$

$$= MPODC_f \cdot \{\overline{MPODC_f} + ODC_f^h + \overline{(f \oplus new\_f)}\}$$

$$= ODC_f^h + MPODC_f \cdot \overline{(f \oplus new\_f)}$$

This proves the claim of this lemma.

■

This lemma suggests a method for computing the observability don't care conditions for node g (Figure 6.6) that guarantees that changes in g will not increase the switching activity of transitive fanout nodes of f. Using the result of this lemma, the following theorem provides a don't care computation technique that guarantees a monotonic reduction in the power consumption of all nodes in the network.

Theorem 6.3 *Consider nodes g, f and h (figure 6.6). Assume f has been optimized using $MPODC_f$ resulting in new_f where new_f is the global function of node f after optimization. $PMPODC_g$, Propagated Monotone Power Relevant Observability Don't Care for node f as defined here, gives the maximal set of don't care conditions that when used in optimizing node g, guarantees that any changes in the function of g does not increase the current switching activity of node f or any node in its transitive fanout cone.*

$$PMPODC_f = \begin{cases} \overline{new\_f} \cdot (ODC_f^h + \overline{f} \cdot MPODC_f) & sp(f) > 0.5 \\ new\_f \cdot (ODC_f^h + f \cdot MPODC_f) & sp(f) \leq 0.5 \end{cases} \quad (6.14)$$

Proof: The case for $sp(f) > 0.5$ is proved first. Proof for the second part is similar.

Since the signal probability of f is greater than 0.5, we need to propagate part of

$MPODC_g$ such that the signal probability of f does not decrease for any changes in function of g. We also need to use part of $MPODC_f$ that guarantees that changes in g do not result in an increase in the power consumption of nodes in the transitive fanout of f. Using theorem 6.1 and lemma 6.7 we can write:

$$PMPODC_f = \overline{new\_f} \cdot MODC_f$$

Using equation 6.13, expanding the XOR term, and simplifying the expression proves the claim of this theorem.

■

Given a node g and its fanout f, $MPODC_g$ the monotone power relevant observability don't care for node g (which is used to optimize the function of g) is calculated using the following equation:

$$MPODC_g = ODC_g^f + PMPODC_f \tag{6.15}$$

Once $MPODC_g$ is used to find new_g (the new function for node g), $PMPODC_g$ (propagated monotone power relevant observability don't care for g) is calculated using the procedure shown in figure 6.8 and stored at node g for computing the don't care conditions for fanins of g.

The advantage of using the techniques proposed in this section is that it guarantees that local optimizations not only improve the local power, but also work on

```
1: function find_propagated_monotone_power_odc(f,new_f,ODC^h_f, MPODC_f)
2:   f is the node function before optimization
3:   new_f is the node function after it is optimized
4:   ODC^h_f is the observability don't care for f at the output h
5:   MPODC_f is the monotone power relevant ODC for f
6:    begin
7:       if (p(new_f) > 0.5 ) then
8:             PMPODC_f = new_f . (ODC^h_f + f̄ . MPODC_f)
9:       else
10:            PMPODC_f = new_f .  (ODC^h_f + f . MPODC_f)
11:      end if
12:   end
```

Figure 6.8 Don't care filter for monotonic decrease in current switching activity

reducing the switching activity of other nodes in the network when possible. These filters, however, result in a smaller don't care sets while optimizing the local function of a node, and this means that it may not be possible to obtain as much reduction in the power contribution of the node to the network power if the power don't care filters were not being used. Therefore, care should be taken in using these filters in order to keep a balance between the available don't care for local optimizations and the don't care being used to guarantee an overall reduction in the global power due to any local changes.

6.1.6 Computing power relevant don't cares in DAGs

The approach used to analyze a tree network is directly used to analyze a general Boolean network. However, since the observability don't care relations for trees do not hold in a general DAG, the number of don't care regions for a node is larger than the number of don't cares regions for trees. The upper bound on the number of don't care regions for a node g in a DAG is given by $2 \cdot 3^k$ where k is the number of transitive fanout nodes of g. This for example is possible if node g has k immediate fanout nodes.

The don't care used for optimizing node g in a DAG is calculated using the following equation.

$$PODC_g = \prod_{f_i \in \text{fanouts(g)}} PODC_g(f_i) \tag{6.16}$$

where $PODC_g(f_i)$ gives the power relevant observability conditions for node g through fanout edge f_i and is computed using equation 6.12. $MPODC_g(f_i)$, computed using equation 6.15, is used to compute $MPODC_g$. However, $PMPODC_{fi}$ for each fanout f_i of g is computed using the following equation derived by extending lemma 6.7 and theorem 6.3.

$$PMPODC_{fi} = \begin{cases} \overline{\text{new\_f}_i} \cdot \left(\left\langle \prod_{h_{ij} \in \text{fanouts(fi)}} ODC_{fi}^{hj} \right\rangle + \overline{f_i} \cdot MPODC_{fi} \right) & sp(f) > 0.5 \\ \text{new\_f}_i \cdot \left(\left\langle \prod_{h_{ij} \in \text{fanouts(fi)}} ODC_{fi}^{hj} \right\rangle + f_i \cdot MPODC_{fi} \right) & sp(f) \le 0.5 \end{cases} \tag{6.17}$$

Note that equations 6.12 or 6.15 provide maximal power relevant don't care sets for tree networks. This maximality does not, however, hold when equation 6.16 is used [15].

The analysis presented in this section computes the maximal set of power relevant observability don't cares for nodes in a tree network. In general it is desired to compute power relevant observability don't care sets that can be used to optimize all nodes without recomputing the don't care sets.

The equations for computing the power relevant observability don't cares have been derived by modifying the part of don't care that is propagated in the network (the second term in equation 6.3) without making any assumption about the first part in this equation, which gives observability conditions for immediate fanout nodes. The operations to generate compatible don't care sets, however, modify the first term in equation 6.3 without making any assumptions on the don't care being propagated. The equations for computing power relevant don't cares are easily extended to generate compatible (in the sense that they can be used to optimize nodes without recomputing the don't care sets) by replacing ODC^f_g in equations 6.12 and 6.15 with that computed in equation 6.4.

The procedures presented in this section provide techniques for calculating the set of _maximal compatible power relevant local don't care_ or _maximal compatible monotone power relevant local don't care_ for nodes in the network, which is then used to optimize the local function of nodes without any concern on degrading the global power consumption of the network. The following section presents a new approach for including the *SDC* in the don't care of the function being optimized.

6.1.7 Power relevant satisfiability don't cares for power optimization

In a Boolean network, some combinations for the values of the internal nodes are not possible no matter what input vector is applied at the primary inputs of the circuit. If a network has n primary input nodes and m internal nodes, satisfiability don't care conditions (*SDC*) for a network contain all impossible combinations in the space of B^{n+m}. The contribution of each node in the network to *SDC* of network is given in definition 6.2. In this sense, *SDC* due to a node g is defined as $g \oplus F(g)$ where g is the variable at the output of node g and $F(g)$ is the function of node g in terms of its immediate fanins. Satisfiability don't cares are usually used to substitute a new variable into a function if this substitution results in a lower cost implementation. The following example illustrates this idea.

Example 6.3:

Consider the Boolean network with functions *f* and *g* where:

$$f = a.b + \bar{a}.\bar{b}$$

$$g = \bar{a}.b + a.\bar{b}$$

While optimizing node *f*, *SDC* due to node *g* can be included in the don't care set of *f*. *SDC* due to node *g* is given as $g \oplus (\bar{a}.b + a.\bar{b})$. Optimizing function *f* using this *SDC* will result in the following form for *f*:

$$f = \bar{g}$$

where *g* is successfully substituted in the function of *f*.

While optimizing a function *f* in the network, a subset of *SDC* is used for nodes that, with high probability, may be substituted into *f*. In [15] a method for selecting a subset of *SDC* is presented where only *SDC* of nodes whose immediate support is a subset of the immediate support of the node being optimized is considered.

It has been shown [12] that using *SDC* due to any node in the network for simplifying the function of a node *f* does not result in changing the global function of *f* or any other node in the network. This means that using *SDC* does not change the signal probability or switching activity of any of the nodes in the network. Therefore, *SDC* may freely be used to optimize the function of nodes without concern that switching activities may increase. Successful use of *SDC* may, however, result in using a new variable *v* in the function of the node being optimized. This results in a load increase at the output of node n_v generating variable *v*. If the switching activity of the new variable is high, then increasing load on this variable may result in an unexpected increase in the power consumption of the network. It is, therefore, important to take into account the switching activity of nodes that are being considered for possible substitution in the function of node that is being optimized.

The approach presented in [15] for using *SDC* consists of identifying a set of nodes that with high probability may be substituted in node *f* being optimized. These nodes are identified as all nodes whose immediate support is a subset of the immediate support of node *f*. The *SDC* due to each of these candidate nodes is then included in the don't care of *f* (see example 6.3). For power optimization, the following approach is used for finding the set of candidate nodes whose *SDC* will be included in the don't care of a node *f*. Figure 6.8 shows the transitive fanin and fanout nodes of node *f* in a Boolean network. *SDC* of nodes in the transitive fanout of node *f* cannot be included in the don't care of *f*. All other nodes in the network may be substituted into *f* if their primary input support is a subset of the primary input support for node *f*. Nodes *g*, *m*, *n* and *p* show such candidate nodes. In order to select a set of candidate nodes, we first find all nodes that are not in the transitive fanout of f and whose primary input support is a subset of the primary input support of f. Among these nodes we select nodes whose switching activity is below a user defined threshold value.

The *SDC* due to a node g (see figure 6.8), whose immediate support is a subset of the immediate support for node f, is easily included as $g \oplus F(g)$. In order to include the *SDC* due to nodes that do not share the same immediate support as f (node m in figure 6.8), it is necessary to include the *SDC* for all nodes that are in the transitive fanin cone of f, and transitive fanin cone of m. This is necessary since successful substitution of m into f requires that the unsatisfiable conditions relating the values at the output of node m, and all immediate fanins of f, be known. The first problem with this approach is that including the *SDC* for all these nodes will result in also considering them for possible substitution into f while f is being optimized, and this may not be desirable if these nodes have a high switching activity. The second drawback is that this operation may prove to be expensive when the number of nodes in the transitive fanin cones of f and m are large. An alternative approach for including *SDC* of nodes that do not share the same immediate support as f is proposed here by observing that given a Boolean network with n primary inputs and m internal nodes, the range of B^n (space of primary inputs) onto B^{n+m} (the space of all nodes in the network), gives the set of all satisfiable conditions for the m input nodes.

Assume an internal node f with fanins $\{i_1, ..., i_l\}$ *is* being optimized while considering possible substitution of nodes $\{n_1, ..., n_k\}$ into the function of f. The complement of the range of the space of primary inputs by function $F=\{i_1, ..., i_l, n_1, ..., n_k\}$ gives all unsatisfiable conditions in the space of function F which is used as the *SDC* set while optimizing node f. Using this technique, it is no longer necessary to include the *SDC* due to nodes which are not good candidates for substitution into function of node f.

The procedures presented in this section compute a set of local don't care for the function that is being optimized. This local don't care guarantees that global power is not degraded while the node is being optimized, and also allows for expressing the node function using a new variable which may potentially result in a lower power consumption. In the next section techniques using minimal literal and variable supports are presented to optimize the local function of a node for low power.

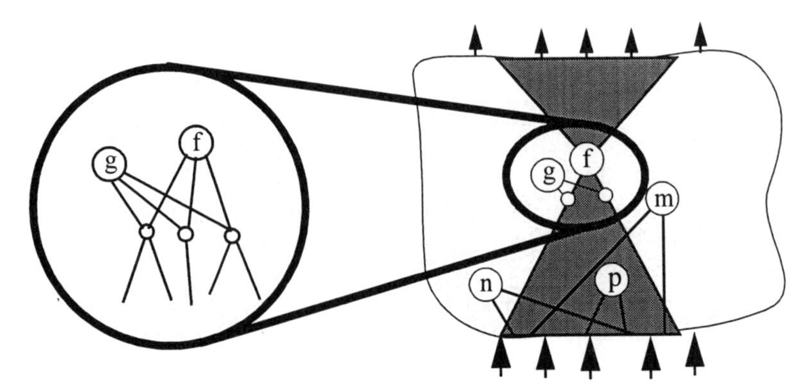

Figure 6.9 Candidate nodes for SDC computation

6.2 Node Minimization using Minimal Variable Supports

The goal of node function minimization for power is to minimize the power contribution of the node to the overall power consumption of the network. This requires that the combination of the node power at the input and output, as well the estimate for the internal power of the node, be minimized. This section presents a method for minimizing the power of the node by reducing load on high activity nodes of the network. We first present a more efficient method for computing the set of minimal literal and variable supports of the nodes in the network. The set of minimal supports gives the flexibility in implementing the node function using different sets of variables. We then propose techniques for selecting a node support which will lead to maximal reduction in the node input power.

6.2.1 Minimal supports of functions

6.2.1.1 Minimal Literal Supports

Given an incompletely specified function *ff*, it is often possible to implement *ff* using different sets of literals. For example let $F = a.b$ and $D_F = a \oplus b$. This function can be simplified to $F = a$ or $F = b$. Then set $\{\{a\}, \{b\}\}$ is the set of all minimal literal supports for node F.

The problem of finding the minimal literal support of a function is stated as follows.

Problem: *Given $C^1=\{C^1{}_0, C^1{}_1,\ldots, C^1{}_g\}$, the cover of the on-set and $C^0=\{C^0{}_0, C^0{}_1,\ldots, C^0{}_h\}$, the cover of the off-set of a function $F(x_1, x_2,\ldots x_n) \in R^n$, find the set all minimal literal supports of the function F.*

The following procedure for finding the set of all minimal dependence sets is presented in [7]. Each of these sets gives the set of literals necessary for implementing the function.

Algorithm 1:

- The problem is first transformed into the R^{2n} space. This is done by replacing the complement of variable x_i in the on-set with a new variable x_{i+n}. The positive occurrences of variable x_i in the cubes of the off-set is replaced with the complement of the new variable x_{i+n}. A new function $f(x_1, x_2,\ldots x_{2n}) \in R^{2n}$ is obtained with cover $c^1=\{c^1{}_0, c^1{}_1,\ldots, c^1{}_g\}$ for the on-set and cover $c^0=\{c^0{}_0, c^0{}_1,\ldots, c^0{}_h\}$ for the offset.
- For each cube $c^1{}_i$ of the on-set and $c^0{}_j$ of the off-set of function f define:

$$H_{ij} = \sum \langle x_k | x_k \supseteq c_i^1, \bar{x}_i \supseteq c_j^0 \rangle . \tag{6.18}$$

  This function corresponds to the union of the set of literals where the presence of each of literal will result in an empty intersection between cubes $c^1{}_i$ and $c^0{}_j$.
- Define function H as:

$$H = \prod_{\substack{i \in 1\ldots g \\ j \in 1\ldots h}} H_{ij} .$$

  The cubes of function H give the minimal support sets of function f in R^{2n}.
- The set of variables in R^{2n} is transformed back to R^n by replacing

variables $n+1$ through $2n$ with the complement of the variables 1 through n. Performing this operation on functions f and H will result in functions F and the minimal literal support for function F.

■

This procedure is explained as follows. H_{ij} for cubes c^1_i and c^0_j gives the set of literals, one of which need to remain lowered in c^1_i in order for this cube not to intersect cube c^0_j of the off-set. This is represented as a conjunctive term. The intersection of all H_{ij} gives all the conditions such that none of the cubes of on-set intersect any of the cubes in the off-set. The reason the problem is transformed into R^{2n} space is that each variable in the R^{2n} space represents the logical event that a positive or negative literal of the function inputs remain lowered. This means that if a minimal literal support requires both positive and negative phases of a variable, then the cube representing this support will not be removed by using the relation $x.\bar{x} = 0$. Note that H is a unate function in all the variables in the R^{2n} space and, therefore, has a unique minimum form representation which consists of all prime implicants of H. Also note that if function F is positive (negative) unate in a variable v, then it is not necessary to introduce a new variable for the negative (positive) literal of variable v. Therefore, if a Function F has n inputs and is unate in m variables, it is sufficient to transform the problem into R^{2n-m} space. The prime implicant cubes of H correspond to the set of minimal literals that need to remain lowered for the on-set not to intersect the off-set.

Once the set of all minimal literal supports of a function has been computed, a literal support is selected to implement the function. This problem is stated as follows:

Problem: *Given* $C^1=\{C^1{}_0, C^1{}_1, \ldots, C^1{}_g\}$, *the cover of the on-set for a function* $F(x_1, x_2, \ldots x_n) \in R^n$, *and a set of minimal literal support given as a set of literals* $MLS_F=\{lit_1, lit_2, \ldots, lit_k\}$, *find the minimal irredundant form of the function F.*

The solution to this problem is obtained by raising all literals in on-set of F which are not a member of set MLS_F [7]. The following example shows how these algorithms are used.

Example 6.4:

Consider the following incompletely specified function F where $g=2$ and $h=1$:

$$on\text{-}set(F) = x_1x_2x_3x_4 + x_1x_2x_3x_4$$

$$off\text{-}set(F) = x_1x_2x_3x_4$$

by setting: $x_5=x_1$, $x_6=x_2$, $x_7=x_3$, $x_8=x_4$ we will have:

$$on\text{-}set(f) = x_5x_6x_3x_4 + x_5x_2x_3x_8$$

$$off\text{-}set(f) = x_1x_6x_3x_8$$

Therefore:

$$H_{11} = x_6 + x_3 \qquad H_{21} = x_8 + x_3 \qquad H = x_6.x_8 + x_3$$

The set of all minimal literal supports of F is given as:

$$\{x_2.x_4\}, \{x_3\}$$

This means that two possible implementation of F are:

$$F = x_3 \qquad or \qquad F = x_2 + x_4$$

6.2.1.2 Reduced Off-sets and Minimal Literal Supports

The method described in [7] for computing the set of all minimal literal supports requires that a cover of the on-set and off-set of the function be computed. The off-set has to be computed by complementing the union of on-set and don't care of the function. This operation is, in general, computation expensive and the resulting off-set might have an exponential size. An example of this function is the Achilles Heel function which has n terms in the cover of on-set and 3^n terms in the cover of off-set. Therefore, it is desirable to compute the set of all minimal literal supports without computing the off-set of the function. This section presents a method for computing the set of all minimal literal supports of a function without computing the off-set by using the ideas behind reduced off-sets.

Reduced off-sets are introduced by observing that some minterms of the on-set or don't care cannot be used to expand a cube of the on-set. Assume $p=\overline{a}.\overline{b}$ and the complete off-set is $a \oplus b$. Then the reduced off-set of p is $(a+b)$, which is all that is needed to expand p.

Definition 6.9 *(Malik [11])Given a cube p of a function f, R_p, the reduced off-set of function f with respect to cube p is obtained by including all minterms of the on-set that cannot be used to expand p, in the off-set of the function.*

It is also shown in [11] that the reduced off-set of a cube is a unate function and therefore, has a unique minimal representation in the SOP form. A procedure is presented in [11] for computing the reduced off-set of each cube in the function where reduced off-sets are computed without computing the complete off-set of the function.

The following theorem shows how reduced off-sets are used to compute the set of all minimal literal supports of a function.

Theorem 6.4*Given a cube $c^l{}_r$ of the cover of the on-set of an incompletely specified* *function and R_r corresponding reduced off-sets for $c^1{}_r$, H_r, the set of all minimal literals supports of $c^1{}_r$ is given by:*

$$H_r = \overline{R_r} \tag{6.19}$$

Proof: Given $\{c^0{}_1,..c^0{}_h\}$, a cover of the off-set for f, define:

$$H_r = \prod_{j=1}^{h} H_{rj} \tag{6.20}$$

By definition:

$$H_r = \prod_{j=1}^{h}\sum_{k}\langle x_k | x_k \supseteq c_r^1, \overline{x_k} \supseteq c_j^0\rangle$$

$$\overline{H_r} = \sum_{j=1}^{h}\prod_{k}\langle \bar{x}_k | x_k \supseteq c_r^1, \overline{x_k} \supseteq c_j^0\rangle = R_r$$

The last equation is by definition equal to the reduced-off-set for cube $c^1{}_r$. Note that the reduced off-set for a cube p of the on-set is unate in all variables. Therefore, it is not necessary to transform the problem into R^{2n} space when finding the support for a single cube of the on-set.

■

Theorem 6.5*Given $F = \{c^1{}_1, c^1{}_2, ..., c^1{}_g\}$ a cover of the on-set and $\{R_1, R_2, ...R_g\}$ the set of corresponding reduced off-sets for an incompletely specified logic function $f(x_1,...x_n)$, the set of all minimal literals supports of f is given by:*

$$H = \prod_{r=1}^{g}\overline{RR_r}$$

where RR_r is obtained by transforming R_r into R^{2n} as described in algorithm 1.

Proof: By definition:

$$H = \prod_{r=1}^{g} H_r.$$

where H_r is as defined in equation 6.20. Theorem 6.4 also showed that $H_r = \overline{R_r}$. This

```
1: function Generate_MLS(F)
2:    F is a Boolean function with cubes (c_1,...c_g)
3:    begin
4:        SupBar = 0
5:            foreach ( cube c_r of the cover) do
6:            R_r = findReducedOffset(F, c_r)
7:            RR_r = transformToNewSpace(R_r)
8:            SupBar = SupBar + RR_r
9:        endfor
10:       Support = NOT(SupBar)
11:       return Support
12:   end
```

Figure 6.10 Computing minimal support using reduced off-sets

means that H can be computed by taking the intersection of the minimal literal supports of each cube of the on-set. The problem, however, is that even though each R_r is unate in all variables, the set of all reduced off-sets are not unate with respect to the same sense of the variables. For example, R_i may be positive unate with respect to variable v and R_j may be negative unate with respect to the same variable. This means that it is necessary to transform each reduced off-set to the R^{2n} space before the intersection of all minimal literal supports has been computed.

■

This approach greatly reduces the complexity of computing the set of all minimal literals supports of an incompletely specified function when the size of don't care set is large. Using this theorem, the procedure shown in figure 6.10 presents an algorithm for generating the set of all minimal literal supports of an incompletely specified function.

6.2.1.3 Minimal Variable Supports

The procedure in section 6.2.1.2 computes the set of all minimal literal supports of a function. For some optimization procedures, it may not be necessary to differentiate between the positive and negative literals of a variable. This means that it is sufficient to compute the set of all minimal variable supports of the function.

The advantage of computing the variable support is that it is no longer necessary to transform the problem into R^{2n} space where n is the number of variables. The following theorem provides a method for computing the set of all minimal variable supports.

Theorem 6.6 *Given $F = \{c^1_1, c^1_2, \ldots, c^1_g\}$ a cover of the on-set and $\{R_1, R_2, \ldots R_g\}$ the set of corresponding reduced off-sets for an incompletely specified logic function $f(x_1, \ldots x_n)$, the set of all minimal literals supports of f is given by:*

$$H = \prod_{r=1}^{g} \overline{\hat{R}_r} \tag{6.21}$$

where $\hat{R}_r$ is obtained by replacing all positive literals in R_r with negative literals.

Proof: The reason that in theorem 6.5, reduced off-sets are transformed into R^{2n} space for literal support computation is that if a literal support has both the positive and negative literals of a function, then the Boolean operations defined for finding the set of minimal literal supports will effectively remove any such support from the solution since a support is represented as a cube containing both positive and negative occurrence of the literal. During variable support computation, each reduced off-set R_r gives the set of minimal literal supports of cube c^1_r. By making all literals negative, then all reduced off-sets are negative unate in all variables which means that it is no longer necessary to transform the problem into R^{2n} space. The literals are also changed into negative literals so that, after complementing the reduced off-sets, the minimal variable supports of the function are represented as positive literals.

■

6.2.2 Node minimization using minimal variable supports

Once the set of minimal variable supports for a function is computed, a decision is made as to which set of variables to use for implementing the function. A simple cost function for minimizing power consumption is to count the number of variables in the variable support. The drawback with this cost function is that it does not consider the switching activity of fanin variables that constitute the support variables. A better cost function is to choose a variable support where the sum of the switching activity for all the variables in the support is minimum. We refer to this procedure as the *minimal switching activity support* procedure. Once the new variable support for a node is determined, the new function of the node is computed by dropping variables not in the support.

When a variable support is selected for the function, a part of the don't care is assigned to eliminate the variables not in the selected support of the node. This operation results in a new function f_{new}. However, a subset of the don't care can still be used to minimize the cover of f_{new}. This subset of the don't care is called reduced don't care $dc_{reduced}$.

Theorem 6.7 *Given a cube v representing the variables removed from the on-set of a function f and dc, don't care for function f, $dc_{reduced}$ the reduced don't care for f, as given below, is the maximal set of don't care that can be used to optimize f without including variables from v in f.*

$$dc_{reduced} = C_v(dc)$$

where v is the bit-wise complement of v.

Proof: Given a function F and a set of variables v, then $C_v(F)$ gives the largest function contained in F which is independent of variables in the set v. The proof for this theorem follows from the observation that $C_v(dc)$ gives the maximal set of don't cares contained in dc which is independent of variables in v.

■

Figure 6.11.a shows the on-set and don't care for a function f. Figure 6.11.b shows the don't care assignment which is used to eliminate variable x from the support to obtain f_{new} and figure 6.11.c shows the reduced don't care for function f after variable x is eliminated from the k-map. Using reduced don't care for this function, one product term is removed from the on-set of function f. Given a function f, its don't care set dc and a variable support v, the procedure in figure 6.12 is used to optimize the power consumption of the function.

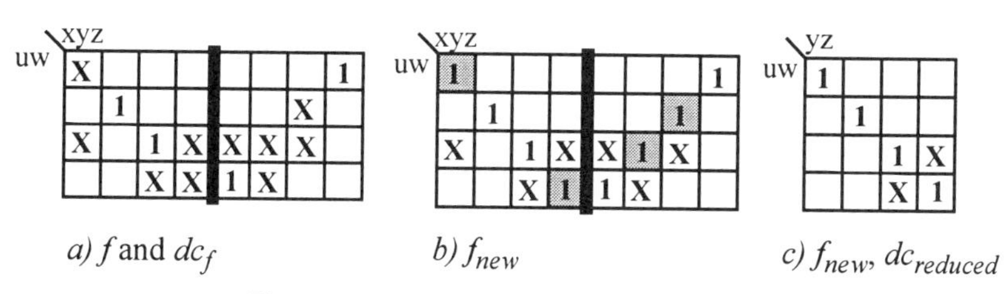

Figure 6.11 Using reduced don't care

The given procedure will provide a low area implementation which has the lowest sum of switching activities on the immediate fanins of the node. It is, however, possible for a variable support with a higher switching activity support cost to have a smaller factored form and hence have a lower power cost. In order to select a variable support which also reduces the node's power estimate as much as possible, we compare the power estimate for the node implementation of the k lowest cost variable supports where k is a user defined parameter. Note that node power cost function in the factored form also takes into account the output load and switching activity. Therefore, by selecting the lowest power cost implementation, a solution is selected which minimizes all factors contributing to the network power consumption. Given a node function f and its don't care dc, the procedure in figure 6.13 is used to select the lowest power cost implementation of the function.

```
1: function node_fanin_optimize(f, dc, v)
2:   f is the function and dc is the don't care of node, v is a variable support;
3:   begin
4:       p= cube representing eliminated variables
5:       f_reduced = function_elim_variables(f, p)
6:       dc_reduced = C_p(dc)
7:       f_new = espresso(f_reduced, dc_reduced)
8:       return f_new
9:   end
```

Figure 6.12 Node minimization using variable support v

```
1: function node_power_optimize(f, dc)
2:   f is the function and dc is the don't care of node;
3:   begin
4:       f_esp = espresso(f, dc).
5:       V = find_k_minimal_switching_activity_var_sup(f, dc).
6:       foreach v ∈ V do
7:           f_tmp = node_fanin_optimize(f, dc, v)
8:           if( power_cost(f_tmp) < power_cost(f_esp) ) then
9:               f_esp = f_tmp
10:          endif
11:      endfor
12:      return f_new.
13:  end
```

Figure 6.13 Node optimization for power

6.3 Experimental Results

The procedures presented in this chapter are implemented in a program called *power_full_simplify*, and the results for benchmarks were compared to those of the *full_simplify* command in the *SIS* package [1].

Each example in the benchmark set is first optimized using *script.ruuged* and then mapped for minimum power. The same example is also optimized using the power script (where *full_simplify* is replaced with *power_full_simplify*), and then mapped for minimum power.

Table 6.2 shows the results of the optimization after running *script.rugged* on benchmark examples, and then mapping for minimum area. Columns 2 and 3 give the area and power of the network right after the *full_simplify* command. The area is given as the number of literals in the factored form, and the power is estimated using the factored form model. Columns 4, 5 and 6 give the network area, delay, and power after technology mapping using library *lib2.genlib* where area, delay, and power are computed using the library parameters. Power is also measured under a zero delay model and randomly set input signal probabilities.

The results in table 6.3 are generated by replacing the *full_simplify* command in *script.rugged* with the *power_full_simplify* command, and then mapping for minimum area. All results are normalized with respect to results in table 6.2. Again, columns 2 and 3 give the area and power estimates before mapping and column 4, 5 and 6 give the area, delay, and power of the mapped networks computed using the library parameters. As results show, on average, power was reduced by %10 with %4 reduction in circuit area. On average, the circuit delay has been increased by %5.

Tables 6.4 shows the runtimes for *full_simplify* and *power_full_simplify* commands. As the results show, *power_full_simplify* is on average 2 times slower than *full_simplify*. In table 6.5, column 2 shows the number of nodes in each example, column 3 shows the average number of inputs considered for each node in a given example, and column 4 gives the average number of minimal variable supports found for each node in the example.

Better results are expected if external don't cares are given for the circuits under consideration. The circuits used in these experiments had no external don't cares; consequently, the *ODC* for these networks were usually small.

Example	Before Mapping		After Mapping		
	Area(factored	Power	Area	Delay	Power
5xpl	113	41.54	118784	31.96	3.60
Z5xpl	116	47.39	122032	34.62	4.19
9sym	211	96.23	226896	21.76	8.21
9symml	186	73.64	205552	22.51	6.10
apex5	777	114.94	934032	41.38	9.87
apex6	743	268.51	814320	25.13	24.76
apex7	245	80.53	266800	20.44	7.44
b12	79	19.28	88160	12.79	1.80
bw	158	38.93	170288	40.79	3.86
clip	132	59.46	147552	21.90	4.83
cps	1237	219.53	1379008	40.67	19.15
des	3462	1077.95	3691584	173.06	95.61
duke2	446	96.13	498800	33.42	8.35
e64	253	34.05	294176	111.02	2.62
ex5	345	89.20	345216	26.86	8.11
example2	331	80.15	362384	19.31	6.98
frg2	886	187.66	875568	42.01	16.34
k2	1135	114.05	1243056	33.13	9.85
misex1	52	15.50	56608	14.28	1.47
misex2	103	22.09	114144	12.43	2.14
pair	1600	504.70	1676432	43.96	43.42
pdc	410	119.24	437552	21.33	10.36
rd84	145	48.60	161472	19.77	4.12
rot	671	204.30	719664	31.54	19.60
spla	648	160.79	756320	25.37	13.30
squar5	56	12.22	59392	23.68	1.08
t481	881	79.68	813856	27.90	6.68
ttt2	219	61.13	232464	17.73	5.29

Table 6.2 Area, delay and power statistics for area script.

Examples	Before Mapping		After Mapping		
	Area(factored)	Power	Area	Delay	Power
5xpl	0.93	0.98	0.86	0.82	0.90
Z5xpl	0.97	0.91	0.95	0.78	0.84
9sym	0.89	1.01	0.83	0.86	0.87
9symml	1.24	1.02	1.15	1.12	0.84
apex5	0.99	0.96	0.96	0.93	0.95
apex6	0.99	0.96	1.00	1.24	0.96
apex7	0.99	0.99	0.97	1.22	0.98
b12	0.96	0.99	0.99	0.92	0.96
bw	1.01	0.87	1.01	0.80	0.90
clip	1.02	0.91	0.94	1.16	0.91
cps	0.99	0.97	1.00	1.17	0.97
des	1.01	1.01	1.02	1.07	0.99
duke2	1.01	1.01	0.99	1.13	0.97
e64	1.00	0.51	0.83	1.16	0.50
ex5	0.99	0.89	0.99	0.97	0.91
example2	0.99	0.96	1.01	1.03	0.97
frg2	0.94	0.86	0.94	1.02	0.86
k2	0.99	0.91	0.98	1.14	0.90
misex1	1.00	0.90	0.96	1.22	0.94
misex2	1.00	0.87	0.97	1.50	0.91
pair	0.99	0.98	0.99	1.04	0.97
pdc	1.00	0.90	1.00	1.09	0.89
rd84	0.97	0.85	0.84	1.13	0.76
rot	0.98	0.94	0.96	1.10	0.95
spla	1.00	0.93	0.99	1.10	0.91
squar5	0.93	0.89	0.86	0.70	0.87
t481	1.00	0.97	1.00	1.01	0.98
ttt2	0.97	0.91	0.90	1.11	0.88
Average	0.99	0.92	0.96	1.05	0.90

Table 6.3 Area, delay and power statistics for power script.

Example	area script	power script	Ratio
5xp1	2.2	7.5	3.41
Z5xp1	2.1	8.4	4.00
9sym	22.6	21.4	0.95
9symml	31.1	36.5	1.17
apex5	25.8	40.3	1.56
apex6	11.3	27.4	2.42
apex7	3.4	6.3	1.85
b12	0.9	2.0	2.22
b9	1.0	3.0	3.00
bw	5.5	7.7	1.40
clip	9.3	29.8	3.20
cps	331.2	375.6	1.13
des	355.1	496.2	1.40
duke2	47.6	87.6	1.84
e64	20.7	35.0	1.69
ex5	13.2	27.9	2.11
example2	4.4	8.5	1.93
frg2	61.6	81.4	1.32
k2	114.0	127.4	1.12
misex1	0.5	1.3	2.60
misex2	0.9	2.1	2.33
pair	69.3	159.2	2.30
pdc	314.1	319.0	1.02
rd84	15.1	21.2	1.40
rot	26.9	69.2	2.57
spla	234.3	318.6	1.36
squar5	0.7	2.4	3.43
t481	215.8	226.1	1.05
ttt2	2.9	8.1	2.79
vda	27.4	38.1	1.39
Average			2.00

Table 6.4 Runtime for the area and power scripts.

Example	Number of	average	average
5xp1	16	6.38	3.06
Z5xp1	16	5.88	3.94
9sym	18	3.72	1.72
apex5	168	6.96	1.68
b12	15	5.27	1.53
bw	34	7.26	2.35
clip	23	6.09	4.83
cps	261	5.95	1.71
duke2	80	6.30	2.10
e64	125	5.39	1.52
ex5	92	6.25	2.86
misex1	10	6.10	1.90
misex2	26	5.81	1.81
pdc	116	6.47	2.19
rd84	23	4.48	3.22
spla	140	6.10	2.25
squar5	9	6.33	3.56
9symml	18	5.50	4.17
apex6	147	6.01	2.32
apex7	65	4.94	1.55
b9	32	4.91	1.94
des	584	6.86	3.32
example2	91	4.75	1.35
frg2	223	6.13	1.94
k2	307	5.88	1.50
pair	294	6.59	1.53
rot	168	5.01	1.36
t481	290	4.78	1.41
ttt2	48	5.50	2.29
vda	165	5.89	1.48
Avg		5.78	2.28

Table 6.5 Average number of node variable supports for each example.

References

[1] "SIS: A system for sequential circuit synthesis," Report M92/41, UC Berkeley, 1992.

[2] R. K. Brayton, G. D. Hachtel, A. Sangiovanni-Vincentelli. "Multi-Level Logic Synthesis." In proceedings of the *IEEE Tranactions*, 78(2):264-300, February 1990.

[3] R. Brayton, G.D. Hachtel, C. McMullen and A. Sangiovanni-Vincentelli. "Logic Minimization Algorithms for VLSI Synthesis." *Kluwer Academic Publishers*, Boston, 1984.

[4] E. Cerny. "An Approach to Unified Methodology of Combinational Switching Circuits." In *IEEE Transactions on Computers*, 27(8), 1977.

[5] O. Coudert, C. Berthet, and J.C. Madre. "Verification of Sequential Machines based on Symbolic Execution." In proceedings of the *Workshop on Automatic Verification Methods for Finite State Systems*, Grenoble, France, 1989.

[6] M. Damiani, G. De Micheli. "Observability Don't Care Sets and Boolean Relations." In proceedings of the *IEEE International Conference on Computer Aided Design*, pages 502-505, November 1990.

[7] C. Halatsis and N. Gaitanis. "Irredundant normal forms and minimal dependence sets of a boolean function." In *IEEE Transaction on Computers*, pages 1064–1068, November 1978.

[8] J. Hong, R. G. Cain, and D. L. Ostapko. "MINI: A heuristic approach for logic minimization." In *IBM journal of Research and Development*, volume 18, pages 443–458, September 1974.

[9] S. Iman and M. Pedram. "An approach for multi-level logic optimization targeting low power." In *IEEE Transactions on Computer Aided Design*, August 1996.

[10] C. K. Leonard, A.R. Newton. "An Estimation Technique to Guide Low Power Resynthesis Algorithms." In proceedings of the *International Symposium on Low Power Design*, pages 227-232, April 1995.

[11] A. A. Malik, R. K. Brayton, A. R. Newton, and A. L. Sangiovanni-Vincentelli. "A modified approach to two-level logic minimization." In proceedings of the *IEEE International Conference on Computer Aided Design*, Nov. 1988.

[12] P. McGeer and R. K. Brayton. "Consistency and Observability Invariance in Multi-Level Logic Synthesis." In proceedings of the *IEEE International Conference on Computer Aided Design*, 1989.

[13] S. Muroga, Y. Kambayashi, H.C. Lai, and J.N. Culliney. "The Transduction Method - Design of Logic Networks Based on Permissible Functions." In *IEEE Transactions on Computers*, October 1989.

[14] R. Rudell. "Logic Synthesis for VLSI Design." Ph.D. thesis, University of California, Berkeley, 1989.

[15] H. Savoj. "Don't Cares in Multi-Level Network Optimization." PhD thesis, University of California, Berkeley, 1992.

[16] A. A. Shen, A. Ghosh, S. Devadas, and K. Keutzer. "On average power dissipation and random pattern testability of CMOS combinational logic networks." In proceedings of the *IEEE International Conference on Computer Aided Design*, November 1992.

CHAPTER 7

Technology Dependent Optimization for Low Power

Chi-ying Tsui

Assistant Professor

Department of Electrical and Electronic Engineering

Hong Kong University of Science and Technology

Power consumption is proportional to the switched capacitance. Minimizing switching activity will thus reduce the power consumption. In the previous chapters, it is shown that the switching activity of an internal node of a circuit depends on its global function and hence by changing the logic structure of the circuit, we can minimize the switching activity. Given a Boolean specification, the technology independent phase of logic synthesis changes the structure of the logic network to reduce the total switching activity of the circuits. As described in chapter 2, the load model used during technology independent optimization is a normalized load value based on the number of fanouts in the factored form or sum-of-products form. Even though this model was shown to be fairly accurate during technology independent phase of optimizations, a more accurate model using the library gate parameters can be used during the technology dependent phase of optimization.

A specific gate library is selected and targeted during the technology dependent phase. A more accurate gate level model is provided in which the lumped capacitive loading at each node of the circuit can be extracted. Therefore, a more accurate switched capacitance value can be used as the objective function during the technology dependent stage of optimization.

In this chapter, we present algorithms to minimize weighted switching activity during the technology dependent phase of logic synthesis. In particular, we show how technology decomposition and mapping are done to minimize power consumption. During the technology mapping stage, zero delay model is used to calculate the switching activities of the nodes. In this case, we only include the switching due to the changing of stable functional logic values and ignore the switching due to glitches. Switching activity due to glitches is more difficult to model accurately since it is related to the relative arrival time of the different inputs to the gate and also the filtering effect due to the inertia of the gate. A real delay model has to be used if glitches are included in the model. However, use of the real delay has an adverse effect on the dynamic programming approach of the mapping algorithm. In most of the circuits, glitching power forms a relatively small portion of the total power and for circuits that have a lot of glitching power, the total power shows a strong correlation with the power estimation using zero delay model, therefore, zero delay model is used during the technology mapping stage. Further power reduction can be achieved after the technology mapping phase. In this chapter, we will discuss gate re-sizing [2] which is used to reduce power consumption for a mapped circuit. The gate re-sizing techniques replace the gates of a mapped circuit that are not on the critical paths with smaller gates in the cell library to reduce the capacitance of the circuit. Chapter 8 presents techniques for structural optimization after technology mapping. Structural transformation makes local transformation by signal substitution in order to reduce the weighted switching activity. The global functions at the primary outputs are preserved during the transformation.

7.1 Technology Dependent Phase of Logic Synthesis

Technology mapping is the final phase of logic synthesis. After technology independent optimization of a set of Boolean equations, the result must be mapped into a feasible circuit which is optimum for some objective functions. The role of technology dependent phase is to finish the synthesis of the circuit by performing the final gate selection from a particular library.

7.1.1 prior work

The two basic approaches used for technology mapping are:

1. rule-based techniques [7][11][17];
2. graph covering techniques [4][15].

A ruled-based system combines the technology-independent and technology mapping stages. It uses local transformations and hence performs local optimization. A rule-based system is a collection of rules and techniques for selecting when and where to apply a rule to improve the circuit quality. Usually, the rule is some transformation between a target and a replacement. A rule is applied by identifying a portion of the circuit which contains a subgraph isomorphic to the target and replacing the subgraph with the replacement without changing the circuit functionality. The circuit quality (with respective to the objective function) is improved through the iterative application of rules.

The graph covering approach uses directed-acyclic-graph (DAG) covering to tackle technology mapping. Keutzer [15] showed that technology mapping is closely related to the problem of code generation in a software compliers. The problem of code generation in a compiler is to map a set of expressions onto a set of machine instructions for the target machine. Each instruction is decomposed into a directed acyclic graph (DAG) of atomic operations, called a pattern, and has a cost associated with it. The sequence of high-level expressions is also represented by a DAG of atomic operations which is called the subject. The optimum code generation problem is equivalent to finding an optimum cost cover of the subject DAG by the pattern DAG's.

For technology mapping, a set of base functions, such as a two-input NAND gate and an inverter, is chosen first. The input technology-independent optimized Boolean equations are then converted into a graph where each node is restricted to one of the base functions. This graph is called the *"subject graph."* The logic function for each library gate is also represented by a graph where each node is restricted to one of the base functions and is called a *"pattern graph."* The technology mapping problem is viewed as the optimization problem of finding a minimum cost covering of the subject graph by choosing from the collection of pattern graphs for all gates in the library.

The quality of the mapping solution depends largely on the input subject graph. The transformation from the technology-independent optimized Boolean equations to the subject graph is called technology decomposition. Singh et. al. [25] proposed a logic restructuring scheme for technology decomposition such that the depth of the decomposed network is minimized. This decomposed network is a good starting point for technology mapping targeting minimum time delay.

For the covering problem, Keutzer [15] showed that for area optimization, a dynamic programming algorithm [14] generates an optimal solution for tree covering. For DAG covering, the problem is NP-hard. He proposed to reduce the problem to a set of tree-covering problems by partitioning the subject graph into trees, covering each tree optimally, and then piecing the tree-covers into a cover of the subject graph.

Rudell [22] tackled the mapping problem for minimum delay optimization for trees, and the constrained-by-timing area optimization problem based on a binning technique for the pin-load. Chaudhary et al. [6] solved the constrained-by-timing area optimization problem by constructing an area-delay trade-off curve during the covering process, and obtaining the mapping solution from the trade-off curve which has minimum area while satisfying the delay constraints.

Since the gate capacitance is proportional to the gate area, the traditional approach for minimizing the power consumption has been to minimize the total gate area. However, the power consumption also depends on the switching activity of the gates. Indeed, we have to minimize the total weighted switching activity in the circuit i.e. $\sum C_{load}E(switching)$. Therefore, the minimum area solution is in general, different from the minimum power solution. In this chapter, we present techniques to minimize the average power consumption during technology-dependent phase of combinational logic synthesis. The idea is to generate a **NAND** decomposition of the Boolean network with minimum switching activity and then ***hide*** high switching nodes within complex gates.

7.1.2 Calculation of signal and transition probabilities

Estimation of transition probabilities depends on the delay model used. Under a ***zero delay model***, where gate delays are assumed to be zero [10], signal transitions due to glitching are ignored. This tends to underestimate the power consumption. There is, however, a good correlation between the estimated power under a zero delay model and a general delay model [24]. In the following, we assume a zero delay model. We will also discuss the issues regarding the use of a general delay model. In the following, we recap the calculation of signal and transition probabilities under a zero delay model.

For **N_TYPE** dynamic circuits, the output is pre-charged to 1 and hence the transition probability is given by

$$\tau p_o^{1 \to 0} = sp_o^{0} \tag{7.1}$$

where sp_o^{0} is the probability of signal o assumes value 0.

For **P-TYPE** dynamic circuits, the output is pre-discharged to 0 and hence the transition probability is given by

$$\tau p_o^{0 \to 1} = sp_o^1 \tag{7.2}$$

where sp_o^1 is the probability of signal o assumes value 1.

For static circuits, we assume that the present input signal value is independent of the previous value, i.e. the primary inputs are temporal independent. Hence, transition probabilities are given by

$$\tau p_o^{0 \to 1} = sp_o^0 sp_o^1 \tag{7.3}$$

$$p_{o_{1(0 \to 1)}|o_{2(0 \to 1)}} = p_{o_1 = 0|o_2 = 0} p_{o_1 = 1|o_2 = 1} \tag{7.4}$$

where $p_{o_{1(0 \to 1)}|o_{2(0 \to 1)}}$ is the conditional probability that signal o_1 has a *0->1* transition given signal o_2 is making a *0->1* transition, and $p_{o_1 = x|o_2 = y}$ is the conditional probability that signal o_1 assumes x given that signal o_2 assumes y. The signal probability at the output of a node is calculated by first building an Ordered Binary-Decision Diagrams (**OBDD**) [3] corresponding to the global function of the node and then performing a linear traversal of the **OBDD** using the procedure given in [19].

7.2 Low Power Technology Decomposition

It is difficult to come up with a decomposed network which will lead to a minimum power consumption implementation after power-efficient technology mapping is applied since gate loading and mapping information are unknown at this stage. Nevertheless, we have observed that a decomposition scheme which minimizes the sum of the switching activity at the internal nodes of the network is a good starting point for power-efficient technology mapping. We illustrate this point with a simple example (see figure 7.1 a). A four-input **AND** gate can be decomposed into a tree of 2-input **AND** gates in two ways. These two decompositions have different total switching activity. Assuming ***n-type*** dynamic circuit and independent inputs, let $sp^1{}_a = 0.7, sp^1{}_b = 0.3, sp^1{}_c = 0.3$ and $sp^1{}_d = 0.7$. The total switching activity for configurations A and B are 2.317 and 2.464, respectively. Configuration A appears to be better than configuration B since the sum of the switching activity at its internal nodes is smaller. Furthermore, if we assume that the cell library has 2-input and 3-input **AND** gates and all gate loadings are the same, the minimum power mapping with a power value of 2.107 is obtained from configuration A (see figure 7.1 b). In

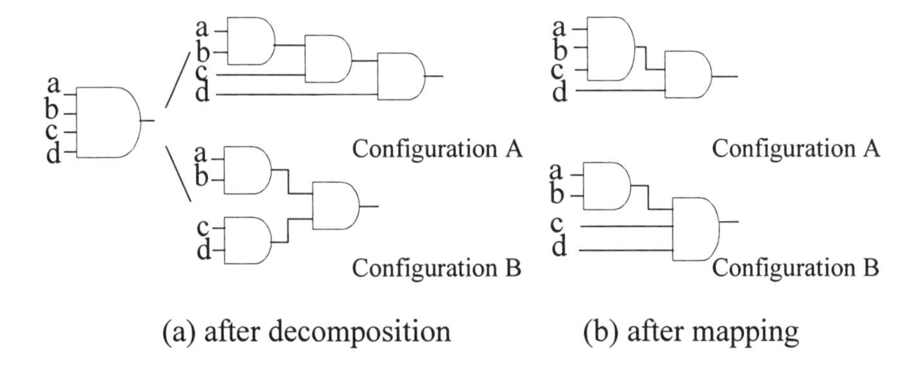

Figure 7.1 Effect of technology decomposition on total switching activity

this example, decomposition with lower switching activity leads to mapping with lower power consumption.

We denote the problem of generating a **NAND**-decomposed network with minimum total switching activity as the **MINPOWER** decomposition. The performance-oriented version of the above problem requires that the increase in the height of the decomposed network (compared to the un-decomposed network) be bounded. We refer to this problem as the **BOUNDED-HEIGHT MINPOWER** decomposition.

7.2.1 Tree decomposition

We describe algorithms for solving the **MINPOWER** decomposition for a fanout-free logic function (i.e. a function that has a tree realization [12]).

The basic algorithm is similar to Huffman's algorithm [13] for constructing a binary tree with minimum average weighted path length. We denote the leaves of a binary tree by $v_1, v_2, \ldots, v_n$, the "path length" from the root to v_i by l_i, and the weight of leaf v_i by w_i. Assuming that the root is at level zero (the highest level), leaf v_i is at level l_i. Given a set of weights w_i, there is a simple *O(n log n)*-time algorithm due to Huffman for constructing a binary tree such that the cost function $\sum_{i=1}^{n} w_i l_i$ is minimum.

(Huffman) Among the n non-negative weights $w_1, w_2, \ldots, w_n$, find the two smallest weights w_1, w_2, say. Replace the two nodes by one node having the weight $W_1 = w_1 + w_2$ and two sons with weights w_1, w_2. Do this recursively for the n-1 weights $W_1, w_3, \ldots, w_n$. The final single node with weight $W_{n-1} = w_1 + w_2 + \ldots + w_n$ is then the root of the binary tree.

It is well known that the resulting binary tree minimizes $\sum_{i=1}^{n} w_i l_i$ over binary trees whose leaves have these weights. We denote this tree as the **MINSUM** tree.

In a more general setting, a *weight combination function F(x,y)* (which is any symmetric function chosen as a binary operator on the weight space U) is used to produce the weight W of internal nodes during tree construction. For each tree T, a *tree cost function* $G(W_1, W_2, \ldots, W_{n-1})$ gives the cost.[1] Parker [20] characterized a wide class of weight combination functions, for which Huffman's algorithm produces optimal trees under corresponding tree cost functions. We give some definitions first and then state Parker's main theorem.

Definition 7.1 *A weight combination function F is quasi-linear if* $F(x, y) = \phi^{-1}(\lambda\phi(x) + \lambda\phi(y))$ *where* λ *is a nonzero constant and* ϕ *is a real-valued invertible function on the weight space U. (Note F is symmetric, and conjugate under* ϕ *to the linear map* $\lambda(x+y)$*.)*

Definition 7.2 *A tree cost function* $G: U^{n-1} \to R$ *is* ***Schur concave*** *if*

$$(x_i - x_j)\left(\frac{\partial G}{\partial x_i} - \frac{\partial G}{\partial x_j}\right) \le 0$$

for all $x_i, x_j \in U$ and $i, j \in 1, \ldots, n-1$.

Theorem 7.1 *If the weight combination function F is quasi-linear and the corresponding function* ϕ *is convex, positive, (or concave, negative) and strictly monotone and* $\lambda \ge 1$, *then the Huffman tree will have the least cost when G is any* ***Schur concave*** *function of the internal node weights.*

1. In Huffman's original paper, $F(x,y) = x + y$ and $G = \sum_{i=1}^{n} W_i$. It is easy to show that $G = \sum_{i=1}^{n} w_i l_i$.

It can be shown that $F(x, y) = max(x, y) + 1$ is a quasi-linear function and its corresponding tree cost function is $G = max_{i=1}^{n-1} W_i$. It is not hard to show that G is equal to $max_{i=1}^{n-1}(w_i + l_i)$ which is the weighted height measure of T.

If $F(x, y)$ does not satisfy the above conditions, then Huffman's algorithm may not produce the optimal solution. We propose the following greedy algorithm to solve the decomposition problem for general weight combination functions.

> For every pair w_i and w_j of the n non-negative weights $w_1, w_2, \ldots, w_n$, compute $F_{ij}(w_i, w_j)$ and store in list L. Find the smallest F_{ij}, say F_{12}. Replace the two nodes by a single node having the weight $W_1 = F_{12}$ and two sons with weight w_1 and w_2. Eliminate all $F_{1k}(w_1, w_k)$ and $F_{2k}(w_2, w_k)$ from L. Compute $F_{1j}(W_1, w_j)$ and insert it into L. Do this recursively for the n-1 weights $W_1, w_3, \ldots, w_n$. The final single node with weight W_{n-1} is then the root of the binary tree.

To solve the **MINPOWER** tree decomposition problem, we must use appropriate weight combination and tree cost functions. We thus distinguish among two cases as follows.

Dynamic Circuits

For ***n-type*** circuits, dynamic gate outputs are pre-charged to 1 and switching occurs when the output changes to 0 during the evaluation phase. For a 2-input **AND** gate composed of a 2-input **NAND** gate and a static inverter (figure 7.2 a), the inverter output is 0 during the pre-charge period and the transition probability (without input signal correlations) is given by

$$W_o = w_{i_1} w_{i_2} \tag{7.5}$$

where W_o and w_{ix} values are $sp_o{}^1$ and $sp_{ix}{}^1$ respectively.

For ***p-type*** circuits, dynamic gate outputs are pre-discharged to 0 and transition occurs when output evaluates to 1. The corresponding formula for the transition probability for a 2-input **AND** gate composed of a 2-input **NAND** gate and a static inverter (figure 7.2b) is then given by

$$W_{\bar{o}} = 1 - (1 - w_{\bar{i}_1})(1 - w_{\bar{i}_2}) \tag{7.6}$$

where the $W_{\bar{o}}$ and $w_{\overline{ix}}$ values are $sp_o{}^0$ and $sp_{ix}{}^0$, respectively.

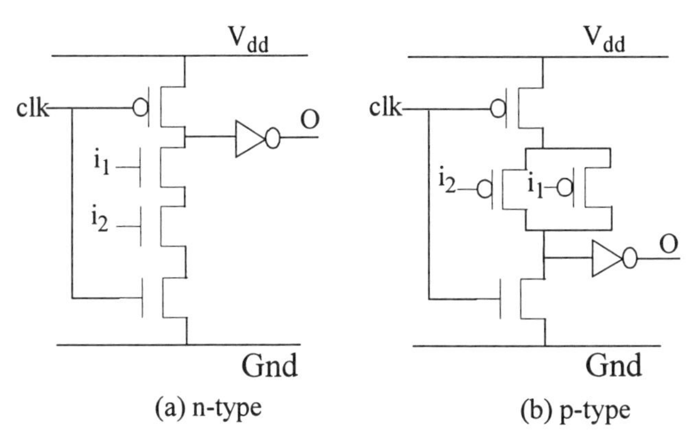

Figure 7.2 Two-input n-type and p-type dynamic AND gates

The weight combination functions $F(w_{i_1}, w_{i_2}) = W_o$ or $F(w_{\bar{i}_1}, w_{\bar{i}_2}) = W_{\bar{o}}$ are used during the **AND** decomposition. The corresponding tree cost function G is given by

$$G = \sum_{i=1}^{n-1} W_i \tag{7.7}$$

Lemma 6.1 *W_o and $W_{\bar{o}}$ given in Eq. (7.5) and Eq. (7.6) above are quasi-linear functions.*

Proof: For Eq. (7.5), since w_i is within the range of [0,1] we can take $\phi(x) = -log(x)$ which is a convex, positive and decreasing function and $\lambda=1$. This shows that W_o is quasi-linear. Similarly, for Eq. (7.6), since $w_{\bar{i}}$ is within the range of [0,1] we can take $\phi(x) = -log(1-x)$ which is a convex, positive and increasing function and $\lambda=1$. This shows that $W_{\bar{o}}$ is quasi-linear.

Lemma 6.2 *G given in Eq. (7.7) above is **Schur concave**.*

Proof: Since $\frac{\partial G}{\partial W_i} = \frac{\partial G}{\partial W_j}$, $(W_i - W_j)\left(\frac{\partial G}{\partial W_i} - \frac{\partial G}{\partial W_j}\right) = 0$, G is ***Schur concave***.

Theorem 7.2 ***MINPOWER** tree decomposition for dynamic **CMOS** circuits with uncorrelated input signals can be solved optimally by Huffman's algorithm using the weight combination functions Eq. (7.5) and Eq. (7.6).*[2]

Proof: Follows from Lemma 6.1 and Lemma 6.2.

If the input signals to the **AND** gate are correlated, Eq. (7.5) and Eq. (7.6) cannot be used as the weight combination function. The transition probabilities for ***n-type*** and ***p-type*** circuits are then given by

$$W_o = w_{i_1} w_{i_2|i_1} \tag{7.8}$$

$$W_{\bar{o}} = 1 - (1 - w_{\bar{i}_1})(1 - w_{\bar{i}_2|i_1}) \tag{7.9}$$

respectively where $w_{i_2|i_1}(w_{\bar{i}_2|i_1})$ is the conditional probability of $i_2(\bar{i}_2)$ given i_1.

Lemma 6.3 *W_o and $W_{\bar{o}}$ given in Eq. (7.8) and Eq. (7.9) are not quasi-linear functions.*

Proof: We present the proof for Eq. (7.8). Proof for the other case is similar. One criterion for a function F to be quasi-linear is increasingness, i.e. $F(u, x) \le F(u, y)$ if $x \le y$ and $F(u, x) \ge F(u, y)$ if $x \ge y$ [20]. However, in Eq. (7.8), we could have $w_{i_2} \le w_{i_3}$, yet $w_{i_2|i_1} > w_{i_3|i_1}$, thus $F(w_{i_1}, w_{i_2}) > F(w_{i_1}, w_{i_3})$. W_o does not satisfy the increasingness property and hence is not quasi-linear.

Since W_o's given in equations Eq. (7.8) and Eq. (7.9) are not quasi-linear, hence we use to solve the decomposition problem.

Static Circuits

For static **CMOS** circuits, we need to minimize the sum of the probabilities for output switching from 0 to 1 and 1 to 0. Thus, the weight combination function W_o for a 2-input **AND** gate is equal to $W_{o_{0\to1}} + W_{o_{1\to0}}$ where

$$W_{o_{0\to1}} = w_{i1_{0\to1}} w_{i2_{0\to1}} + w_{i1_{1\to1}} w_{i2_{0\to1}} + w_{i1_{0\to1}} w_{i2_{1\to1}} \tag{7.10}$$

$$W_{o_{1\to0}} = w_{i1_{1\to1}} w_{i2_{1\to0}} + w_{i1_{1\to0}} w_{i2_{1\to1}} + w_{i1_{1\to0}} w_{i2_{1\to0}} \tag{7.11}$$

If input signals are correlated, conditional probabilities between the input signal transitions have to be used in order to calculate the output transition probability. $W_{o_{0\to1}}$ and $W_{o_{1\to0}}$ are then given by

$$W_{o_{0\to1}} = w_{i1_{0\to1}} w_{i2_{0\to1}|i1_{0\to1}} + w_{i1_{1\to1}} w_{i2_{0\to1}|i1_{1\to1}} + w_{i1_{0\to1}} w_{i2_{1\to1}|i1_{0\to1}} \tag{7.12}$$

[2]. Indeed, a chain-like tree decomposition will be obtained.

$$W_{o_{1\to 0}} = w_{i1_{1\to 1}} w_{i2_{1\to 0}|i1_{1\to 1}} + w_{i1_{1\to 0}} w_{i2_{1\to 1}|i1_{1\to 0}} + w_{i1_{1\to 0}} w_{i2_{1\to 0}|i1_{1\to 0}} \quad (7.13)$$

Note that for static circuits, every internal node is assigned multiple weights (i.e. $w_{i_{0\to 1}}, w_{i_{1\to 0}}, w_{i_{1\to 1}}$). The weight combination uses all these weights through equations (7.10) to Eq. (7.13). This general class of problems cannot be solved using the Huffman's algorithm which applies when each internal node is given a single weight. Therefore, we resort to Eq. .

Under the temporal independence assumption for the gate inputs, Eq. (7.10) and Eq. (7.11) can be replaced by

$$W_{o_{0\to 1}} + W_{o_{1\to 0}} = 2W_{i_1} W_{i_2}(1 - W_{i_1} W_{i_2}) \quad (7.14)$$

while Eq. (7.12) and Eq. (7.13) can be replaced by

$$W_{o_{0\to 1}} + W_{o_{1\to 0}} = 2W_{i_1} W_{i_2|i_1}(1 - W_{i_1} W_{i_2|i_1}) \quad (7.15)$$

The resulting weight combination functions $F(W_{i_1}, W_{i_2})$ are not quasi-linear, and hence is used.

7.2.2 Bounded-height tree decomposition

The objective here, is to construct a **MINPOWER** binary tree for a given list of weights (signal transition probabilities) with the restriction that the height of each weight (defined as $max_i\ l_i$) does not exceed a given integer L. The best known algorithm for solving **bounded-height minsum** problem is an $O(nL)$ algorithm due to Larmore and Hirschberg [16]. Their approach (based on the **package-merge** algorithm) transforms the **bounded-height minsum** tree decomposition problem to an instance of the **coin collector**'s problem.[3] The **package** step in this algorithm uses the Huffman's Algorithm to merge two items at level i to form a new item at level $i+1$. The **merge** step merges the newly formed items with the original items at each level. The following is a summary of the algorithm and details can be found in [16].

[3]. An instance (I,X) of the **coin collector**'s problem is defined as given a set I of m items each of which has a *width* and *weight*, find a subset of S of I whose widths sum to X and has minimum total weight.

Let n be the number of leaf nodes and L be the height bound. Define a node to be an ordered pair(i,l) such that $i \in [1, n]$, which is called the index of the node and $l \in [1, L]$, which is called the level of the node. Define $width(i, l) = 2^{-l}$. If T is a tree, define node_set(T) = $\{(i, l) | (1 \le l \le l_i)\}$ where l_i is the depth of ith leaf of T. Let T_i be the tree we want to decompose with depth no more than L and let each node in the *node_set* be an item which has width less than 1. For each $l \in [1, L]$, the list of nodes with width 2^{-l} is initialized as $((n,l),(n-1,l),...,(1,l))$. The optimal Huffman tree with height bound L is thus equivalent to finding a minimal weight *node_set* of width n-1 which can then be reduced to an instance of the Coin Collector's problem of size nL. An instance(I,X) of the Coin Collector's problem is defined as given a set I of m items each of which has a *width* and *weight*, find a subset of S of I whose widths sum to X and has minimum total weight. The problem is solved by ***Package-Merge*** algorithm which is described by the pseudo-code in figure 7.3.

```
1: Package-Merge Algorithm(I,X)
2:   S = 0
3:   for all d, L_d = list of items having width 2^d, sorted by weight
4:   while X > 0 loop
5:       minwidth = the smallest term in the diadic expansion of X
6:       if I = 0 then
7:              return No solution
8:       else d = the minimum such that L_d is not empty
9:              r = 2^d
10:             if r = minwidth then
11:                return No solution
12:             else if r > minwidth then
13:                Delete minimum weight item from L_d and insert into S
14:                X = X-minwidth
15:              end if
16:             P_{d+1} = PACKAGE(L_d)
17:                (PACKAGE merge items in consecutive pairs in L_d to
                          form a new item in L_{d+1})
18:             L_{d+1} = MERGE(P_{d+1},L_{d+1})
19:      end if
20:  end loop
```

Figure 7.3 Package-Merge Algorithm

The **BOUNDED-HEIGHT MINPOWER** tree decomposition problem for dynamic circuits with uncorrelated input signals can be solved optimally in time *O(nL)* by Larmore-Hirschberg's algorithm using the weight combination functions Eq. (7.5) and Eq. (7.6) since they are quasi-linear and Huffman algorithm can produce the optimal decomposition.

For the general weight combination functions, the Larmore-Hirschberg's algorithm has to be modified as follows:

In the **PACKAGE** step, replace the Huffman's algorithm by . An item at level *i* is obtained by merging the pair of items at level *i-1* which has the minimum weight combination value. The **MERGE** step is unchanged. We therefore perform n^2 weight calculations during the **PACKAGE** step. Using the modified Larmore-Hirschberg's algorithm, the **bounded-height minpower** tree decomposition can be solved heuristically for the general weight combination functions. In the **PACKAGE** step, items are combined and selected according to the weight combination function described in Section 7.2.1. In the **MERGE** step, two item lists are merged and sorted according to the corresponding weight combination function described in Section 7.2.1. The modified algorithm is greedy in nature and does not guarantee optimality.

7.3 Low Power Technology Mapping

The problem of technology mapping for low power can be stated as follows: Given a Boolean network representing a combinational logic circuit optimized by technology independent synthesis procedures and a target library, we bind nodes in the network to gates in the library such that the average power consumption of the final implementation is minimized and timing constraints are satisfied. A successful and efficient solution to the minimum area mapping problem was suggested in [15] and implemented in programs such as DAGON and MIS. The idea is to reduce technology mapping to DAG covering and to approximate DAG covering by a sequence of tree coverings which can be performed optimally using dynamic programming. [27] extended this approach to solve the technology mapping problem minimizing delay (subject to error due to unknown load values during mapping) and the technology mapping problem minimizing area under delay constraints. A near-optimal solution to the latter problem is presented in [6]. Their approach consists of two steps. First, a postorder traversal is used to determine a set of possible arrival times at the root of the tree. Once the user specifies a single required time, a second, preorder traversal is performed to determine a specific technology mapping solution. This scheme is similar to that proposed in [26] in order to solve the optimal orientation problem for a slicing tree of macro-cells. For a NAND-decomposed tree, subject to load calculation errors, this two-step approach finds the minimum area mapping satisfying all delay constraints if such a solution exists.

Technology mapping for low power follows a procedure similar to the above, except that the objective is to minimize the sum over all gates of the weighted switching activity in the mapped network subject to given required time constraint. The approach also consists of two steps. First a postorder traversal is used to determine a set of possible arrival times and accumulated power consumptions at each node of the network. Once the user specifies a single required time, a second preorder traversal starting from the primary outputs is performed to determine the mapping solution that minimizes the average power subject to the required time constraints.

7.3.1 Terminology

The following terminology is borrowed from [6]. Consider a match g at node n of a **NAND**-decomposed tree. The inputs to node n consist of nodes n_i which fanout to node n (that is, $n = n'_1 + n'_2$ if n has two inputs or $n = n'_1$ if n has a single input). The nodes which are covered by match g are denoted by *merged(n,g)*. The nodes which are not in *merged(n,g)* but fanin to *merged(n,g)* are denoted by *inputs(n,g)*. The *mapped-parent(n_i)* is some node n for which there exists a matching gate g such that $n_i \in inputs(n, g)$. Note that given node n and gate g matching at n, *inputs(n,g)* are uniquely determined. However, n_i may have many distinct mapped-parents. Figure 7.4 shows an example of the terminology.

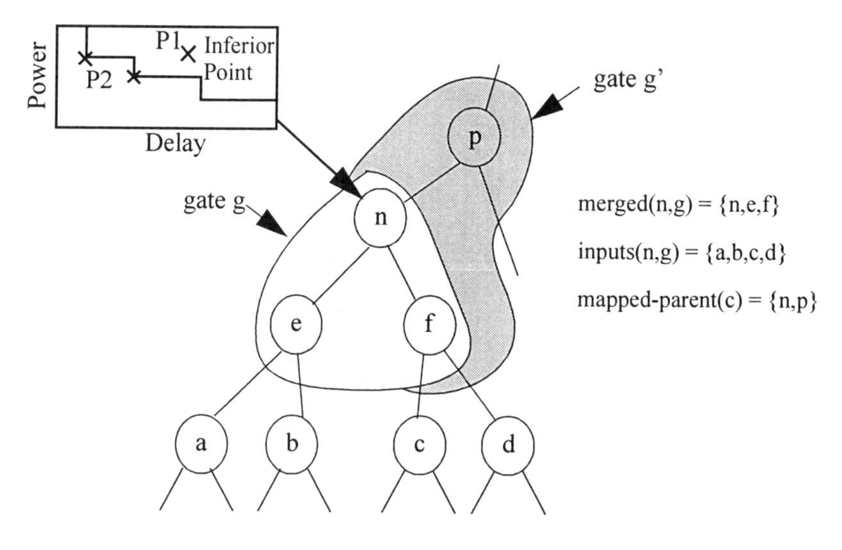

Figure 7.4 Terminology

With each node in the network a power-delay curve is stored. A point on the curve represents the arrival time at the ***output*** of the node and the average power consumed in its mapped transitive fanin cone (but excluding the power consumed at the load driven by the node which has yet to be decided by a later mapping). In addition to the power and delay values, the matching gate and input bindings for the match are also stored with each point on the curve. Points on the curve represent various mapping solutions with different trade-offs between average power and speed. A point (t_1,p_1) is inferior if there exists another point (t_2,p_2) on the curve such that either $p_1 > p_2$ and $t_1 \geq t_2$ or $p_1 \geq p_2$ and $t_1 > t_2$ (see figure 7.5). All inferior points are dropped from the power-delay curve.

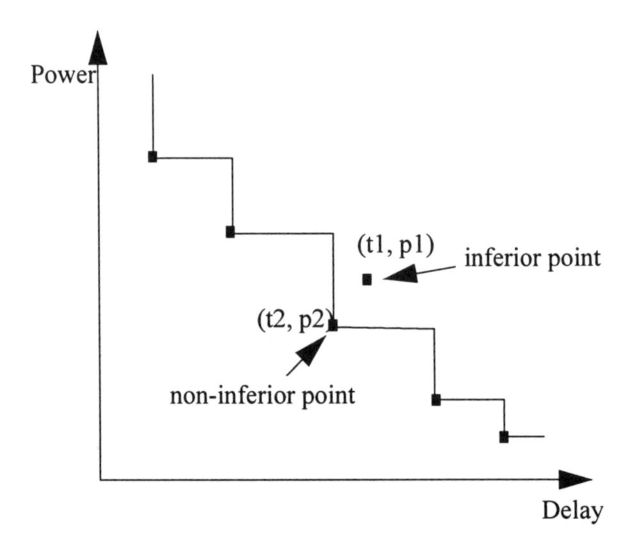

Figure 7.5 Inferior and non-inferior points on a power-delay curve

7.3.2 Arrival time and power cost calculation

The pin-dependent **SIS** library delay model is adopted for the calculation of the arrival time and power consumption. Suppose that gate g has matched at node n, then the output arrival time at n is given by [6]

$$A(n, g, C_n) = max_{n_i \in inputs(n, g)}(\tau_{i,g} + R_{i,g}C_n + A(n_i, g_i, C_{n_i}))$$

where $\tau_{i,g}$ is the **intrinsic** gate delay from input i to output of g, $R_{i,g}$ is the **drive** resistance of g corresponding to a signal transition at input i, C_n is the load capacitance seen at n, $A(n_i, g_i, C_{n_i})$ is the arrival time at input i corresponding to load C_{n_i} seen at that input, and g_i is the best match found at input i. When the gate g matches at n, the fanout of n is not yet mapped and hence the actual load capacitance seen at n (C_n) is still unknown. This is called the **unknown load problem**. A default value is assumed for C_n during the calculation of the arrival time for n. This arrival time has to be recalculated once when the fanout of n is mapped and C_n is known. The details of the timing recalculation is discussed in the next Section.

The input to the technology mapper is a 2-input **NAND(NOR)** decomposed network of the optimized network from the technology independent phase. The signal probability sp_n and the switching activity E_n of each node n of the decomposed network is calculated prior to the mapping. The power cost is thus given by

$$P_{avg}(n, g) = \sum_{j \in inputs(n, g)} \left(0.5 \frac{V_{dd}^2}{T_{cycle}} C_{n_i} E_{n_i} + P_{avg}(n_i, g_i) \right) \qquad (7.16)$$

where E_{n_i} is the expected switching activity at node n_i, $P_{avg}(n_i, g_i)$ is the accumulated power cost at n_i assuming a gate matching g_i.

When calculating the power cost at node n, the power contribution from the output load driven by n is not included. We, however, have the following lemma for the power cost calculation:

Lemma 6.4*Under a zero delay model, the switching activity at node n only depends on the global function of the node in terms of the circuit primary inputs and not its particular implementation. Hence, E_n is independent of the gate matching at n. Furthermore, note that since the power consumption due to output load of n is calculated only when the fanout gate of n is known, (7.16) is* ***not*** *subject to the* ***unknown load problem****. The average power contribution of the n's output load will be included at* ***mapped-parent(n)****. When the mapping reaches a primary output, every point on the power-delay curve has a* ***power*** *value equal to the total average power consumption of the mapped tree minus the power consumption at the primary output load. The power consumption at the primary output load depends on E_n at the output and the load capacitance connected to it which are both* ***independent*** *of the mapping configuration. Hence, it only causes a fixed shift of the curve along the power axis. It does not affect the selection of the optimal point from the power-delay curve.*

7.3.3 Tree mapping

In this section, we focus on tree mapping. Later, we will describe the extension to **DAG** mapping. We specifically describe two tree-traversal operations which are applied to a **NAND**-decomposed tree in order to obtain a technology mapping solution which minimizes the average power consumption while satisfying the timing constraints.

7.3.3.1 Postorder traversal

On the first traversal, we begin at the leaf nodes of the **NAND**-decomposed tree. Since each node n possesses a set of possible arrival time - average power points which are reflected in its power-delay function, the power-delay function at any *mapped-parent*(n) must also reflect these possible arrival time-average power trade-offs. A postorder traversal of the **NAND**-decomposed tree is performed, where for each node n and for each gate g matching at n, a new power-delay function is produced by appropriately merging the power-delay functions at the *inputs(n,g)*. Merging must occur in the common region in order to ensure that the resulting function reflects feasible matches at the *inputs(n,g)*. The power-delay functions for successive gates g matching at n are then merged by applying a ***lower-bound merge*** operation on the corresponding power-delay functions [6]. At a given node n, the resulting power-delay function will describe the arrival time - average power trade-offs in propagating a signal from the tree inputs to the output of n.

7.3.3.2 Preorder traversal

The second traversal begins when the mapping reaches the root (primary output). The user is allowed to select the arrival time - average power trade-off which is most suitable for his application. Given the required time t at the root of the tree, a suitable *(t,p)* point on the power-delay function for the root node is chosen. The gate g matching at the root which corresponds to this point and *inputs(root, g)* are, accordingly, identified. The required times t_i at *inputs(root, g)* are computed from t and g. The preorder traversal resumes at *inputs(root, g)* where t_i is the constraining factor and a matching gate g_i with minimum power consumption satisfying t_i is sought.

7.3.3.3 Timing recalculation

The gate delay is a function of the load it is driving. During the postorder tree traversal, the output of current node n_i, is not mapped hence the load capacitance is unknown (unless, all the gates in the library have identical pin capacitances). At this time the load value is assumed to be that offered by the smallest two-input **NAND** gate in the library. When we come to a node n = *mapped-parent*(n_i) with matching gate g, we know the exact load seen by n_i. This load is equal to the input capacitance of g and is, in general, different from the default load. Therefore, in order to calculate the arrival time at node n, the output arrival times for all nodes in *inputs(n,g)* must be adjusted to account for the change in the load capacitance. Similarly, during the preorder tree traversal, when a gate g is selected to match at n, the load seen by *inputs(n,g)* must be recalculated.

In order to account for this load change (δ_i), the power-delay curves at the inputs have to be appropriately shifted. In particular, since the gate matching g at n_i and giving rise to a point p_j on power-delay curve of n_i is stored with that point, the delay shift is computed as $\delta_i \times R_g$ where R_g is the drive resistance of g.

7.3.3.4 Optimality of the tree mapping algorithm

The following theorem can be stated.

Theorem 6.3 *Under a zero delay model, the tree mapping algorithm finds the minimum power consumption solution for a tree network given delay constraints (subject to the error due to unknown loads during the arrival time calculation).*

Proof: From Lemma 6.4, under a zero delay model, the power consumption calculation is exact and not subject to the ***unknown load problem*** and consequently is similar in nature to the area calculation. The proof of optimality of the tree mapping algorithm is thus similar to that of the area-delay mapping algorithm [6].

7.3.4 DAG mapping

Most of the real circuits are not trees, but general **DAGs**. The problem of mapping a **DAG**, even for the unit fanout model is **NP**-hard [5]. Therefore, we resort to heuristics. During the power-delay curve computation step, nodes are visited in postorder. For each node, we compute the power-delay curve as in the case of trees. However, if the input for a candidate match at the node is coming from a multiple fanout node, we divide the average power contribution of that input by the fanout count of the input node. By reducing the average power contribution, we favor a solution in which multiple fanout nodes are preserved after mapping, which reduces logic duplication and improves the final mapped average power. This heuristic, which permits tree boundary crossing only for nodes with relatively few fanouts, was also adopted by the MIS mapper [8]. During the gate selection step, if we come to a node which has already been mapped, we check if the mapped solution at the node satisfies the timing requirement. If so, we keep the mapping; otherwise, we replace it with another solution from the power-delay curve which satisfies the current timing requirement and has minimum average power.

7.3.5 Experimental results

The technology decomposition and mapping algorithm for low power were applied on a subset of ISCAS-89 and MCNC-91 benchmarks. Static **CMOS** circuits were used in the experiments and all primary inputs were assumed to be pairwise and temporally independent. The delay was calculated using the augmented pin-dependent **SIS** library delay model as described in Section 7.3.2 and an industrial library with 44 gates was used.

Table 7.1 shows the total switching activity of the decomposed networks for different technology decomposition methods. It is shown that the low power technology decomposition reduces the total switching activity in the networks by 5% over the conventional balanced tree decomposition method.

Table 7.2 and Table 7.3 contain the experimental results using different technology decomposition and mapping combinations. All methods have the same starting point, that is, circuits optimized by the **SIS** rugged script [23]. Method A uses area-delay mapping (***ad_map***) algorithm of [6] and methods B to D use power-delay mapping (***pd_map***). Methods A and B take in a **NAND**-decomposed network generated by the conventional balanced tree decomposition algorithm. Method C uses the **minpower** technology decomposition (***minpower_t_decomp***) while method D uses the ***bounded-height minpower*** decomposition (***bh_minpower_t_decomp***).

To see the impact of the ***pd_map*** on the average power consumption, we compare results of methods A and B. It is seen that with ***pd_map***, the power consumption is improved by an average of 15.9% over ***ad_map***. The active cell area is increased by an average of 12.2% while the circuit delay is improved by 2%.

To illustrate the impact of the ***minpower_t_decomp*** on the average power consumption, we compare the results of methods of B and C. The power consumption is improved by an average of 2.5% at the expense of 3.5% increase in area and 2.8% degradation in performance. From B and D, we see that ***bh_minpower_t_decomp*** improves the power consumption by an average of 1.6% over the conventional decomposition at the expense of 0.8% increase in area and 1.8% degradation in performance (Note that ***bh_minpower_t_decomp*** improves performance by an average of 1% over ***minpower_t_decomp***). The best overall result is achieved when ***pd_map*** is applied after the network is decomposed using ***minpower_t_decomp***. The average power consumption is reduced by an average 18% at the expense of 16% increase in area while the circuit delay is unchanged.

Comparing Table 7.1 and Table 7.3, we observe that although the reduction in power consumption after mapping is not directly proportional to the reduction in the switching activity of the network before mapping, networks with lower switching activity often result in a mapped circuit with lower power consumption. We believe the small gain of ***minpower_t_decomp*** is due to the fact that most nodes in the optimized network are relatively simple due to the fast-extract and quick decomposition operations performed prior to the technology decomposition step. Therefore, the ***minpower_t_decomp*** does not have much freedom in improving the power efficiency through **NAND** decomposition.

circuit	balanced decomposition	power eff. decomposition	reduction in switching activity
C1908	672	657	2.23
C432	335	318	5.07
alu2	422	389	7.82
apex7	394	385	2.28
cordic	119	110	7.56
example2	473	457	3.38
pair	2628	2491	5.2
parity	110	110	0
pm1	75	70	6.67
s208	132	128	3.03
s349	207	206	0.48
s382	245	233	4.90
s386	125	109	12.80
s420	247	239	3.24
s400	275	258	6.18
s444	252	245	2.78
s510	410	392	4.39
s641	300	282	6
s713	309	284	8.09
s820	401	367	8.48
s832	385	352	8.57
ttt2	339	327	3.54
x1	499	453	9.22
x3	1224	1192	2.61
x4	598	584	2.34
Average reduction			5.08

Table 7.1 The total switching activity for different decomposition schemes

circuit	***ad_map*** (with balanced decomposition)			***pd_map*** (with balanced decomposition)		
	Power	Area	Delay	Power	Area	Delay
C1908	282.54	6091.4	26.17	256.52	6945.86	25.36
C432	131.18	2278.00	30.73	110.32	2429.64	30.11
alu2	200.67	4266.32	26.05	163.80	5246.88	29.03
apex7	142.02	2734.96	11.69	124.84	2990.64	11.62
cordic	47.97	820.76	6.43	39.87	921.74	6.62
example2	176.77	3822.28	11.30	152.73	4371.04	9.76
pair	1001.3	18380.40	28.20	888.07	21583.8	29.86
parity	52.45	853.40	7.01	41.91	795.60	6.14
pm1	31.11	592.96	5.17	23.77	659.60	5.15
s208	49.06	867.68	8.14	41.16	1007.76	9.99
s349	83.00	1637.44	12.57	71.76	1753.72	13.14
s382	97.26	1878.16	10.42	86.79	2127.72	10.44
s386	64.88	1579.64	11.36	46.69	2031.84	11.71
s420	98.49	1829.88	10.77	88.01	2056.32	11.26
s400	98.39	1785.00	15.23	84.46	2139.96	18.39
s444	101.44	1819.68	10.97	88.82	1993.76	12.24
s510	164.13	2869.60	20.39	141.85	3376.88	15.78
s641	108.22	2137.92	21.12	89.87	2297.04	20.21
s713	108.47	2119.56	21.52	91.06	2264.40	20.01
s820	183.51	3423.12	12.59	146.02	3982.76	11.36
s832	178.53	3450.32	14.30	141.20	3941.28	10.98
ttt2	146.04	2528.92	12.26	122.80	2760.12	12.67
x1	204.66	3683.56	9.38	167.82	4025.60	8.14
x3	485.68	9200.40	13.45	411.14	10509.06	11.50
x4	227.05	4714.44	18.79	194.30	4833.44	15.68

Table 7.2 Technology decomposition and mapping results

circuit	*pd_map* (with ***minpower_t_decomp***)			*pd_map* (with ***bh_minpower_t_decomp***)		
	Power	Area	Delay	Power	Area	Delay
C1908	249.76	7226.36	26.84	254.41	7058.06	25.57
C432	106.39	2506.48	30.32	110.07	2456.84	29.12
alu2	161.28	5451.56	29.04	159.59	5375.40	28.50
apex7	123.89	3123.92	11.93	124.33	3090.94	11.52
cordic	37.65	866.32	8.26	38.80	879.92	7.24
example2	150.00	4522.68	10.21	150.92	4426.12	12.16
pair	867.31	21803.52	29.57	878.25	21896.68	29.75
parity	41.91	795.60	6.14	41.91	795.60	6.14
pm1	22.60	683.40	5.88	22.59	637.84	5.82
s208	40.54	1054.00	9.87	40.68	1054.68	10.42
s349	71.78	1754.40	13.10	72.14	1765.96	13.10
s382	85.09	2188.92	10.38	85.31	2086.24	10.38
s386	45.19	2114.12	11.16	45.18	2036.60	10.49
s420	86.73	2112.7	11.13	87.39	2034.56	11.40
s400	81.06	2290.24	19.40	81.58	2243.32	18.62
s444	88.10	2043.40	12.14	88.46	1999.20	12.14
s510	138.73	3442.8	15.22	138.90	3344.92	15.50
s641	86.93	2512.60	19.69	88.56	2372.52	20.12
s713	87.99	2513.28	21.85	90.18	2337.84	20.10
s820	139.58	4162.28	10.51	141.55	4009.28	10.67
s832	134.25	4223.48	11.33	138.28	4024.92	10.86
ttt2	121.67	2887.28	12.50	122.56	2748.56	12.91
x1	159.14	4293.52	8.13	161.59	4125.56	8.94
x3	406.71	10691.64	14.23	409.77	10643.02	12.73
x4	191.04	4870.84	15.75	190.67	4784.48	15.26

Table 7.3 Technology decomposition and mapping results

It is noteworthy that in most cases, in order to reduce the power consumption, circuit area is increased. To minimize power consumption, ***pd_map*** hides the high switching nodes inside the complex gates and/or reduces the load capacitance of these nodes. In general, these two operations do not produce minimum area mapping. Figure 7.6 shows the minimum area and the minimum power mappings of the ***C17*** circuit, respectively. While the area of the minimum power mapping is larger, the load on the high switching nets is smaller.

To see how power consumption is actually reduced by ***pd_map***, we selected the s832 circuit and plotted the number of nets and the average loading on nets versus the switching activity. Figure 7.7 and 7.8 summarize the results. From figure 7.7, we can see that ***pd_map*** tends to reduce the number of high switching activity nets at the expense of increasing the number of low switching activity nets. From figure 7.8, we can see that for the remaining high switching activity nets, ***pd_map*** tries to reduces the average loading on nets. By doing these, ***pd_map*** minimizes the total weighted switching activity and hence the total power consumption in the circuit.

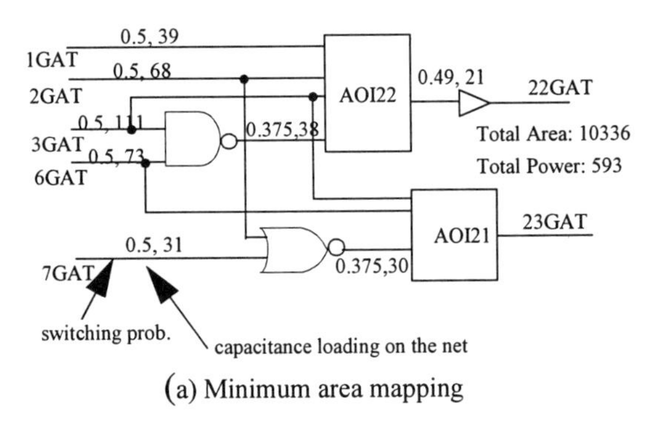

(a) Minimum area mapping

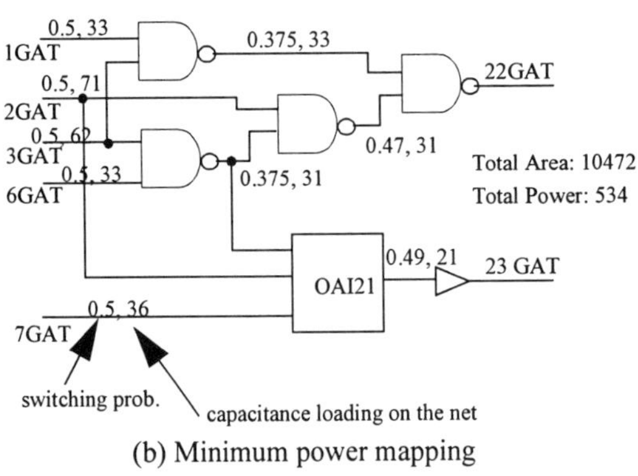

(b) Minimum power mapping

Figure 7.6 Minimum area versus minimum power mappings for C17

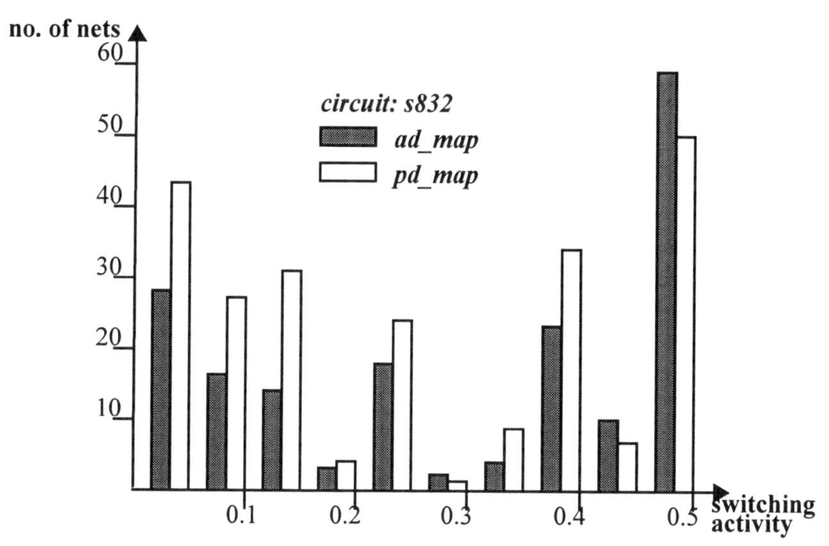

Figure 7.7 Number of nets vs. switching activity for s832 using ***ad_map*** versus ***pd_map***

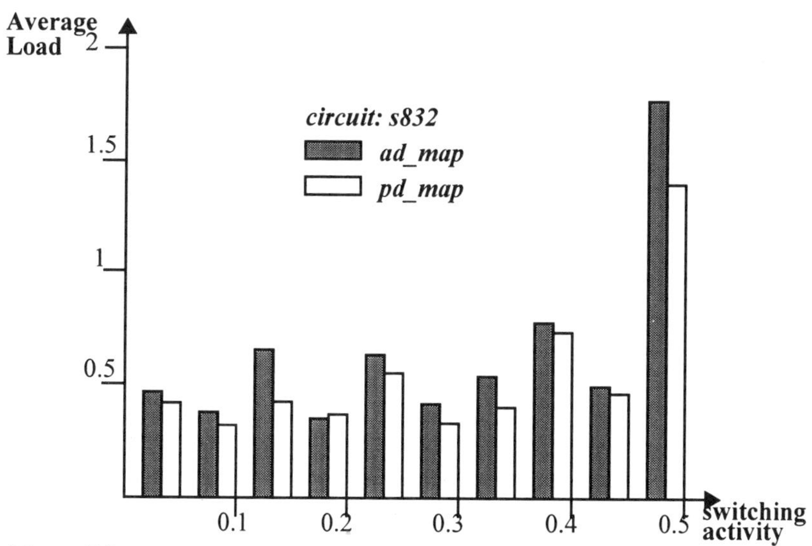

Figure 7.8 Average load per net vs. switching activity for s832 using ***ad_map*** versus ***pd_map***

7.3.6 Discussion on using real delay model

We assumed a zero-delay model which does not account for power consumption due to glitches during the technology mapping phase. This shortcoming can be overcome by using a general delay model such as the **SIS** pin-dependent library delay model when calculating the expected number of transitions. Using a general delay model has, however, several drawbacks. The first is the huge computational effort to calculate the possible glitches for each nodes. In [10], symbolic simulation is used to produce a set of Boolean functions which represent conditions for switching at each gate in the circuit at a specific time instance. This procedure is exact; however it requires large computation time and storage space. A faster method of estimating the switching activity including glitches using the notion of transition probability [19] [28] can be used. The transition probabilities are propagated from the primary inputs to the circuit outputs using a linear time algorithm. Although the later method significantly reduces the computation time and space requirement, it is still costly.

The second drawback is that under a general delay model, the dynamic programming based tree mapping algorithm does **NOT** guarantee an optimum solution even for a tree. The mapping algorithm assumes that the current best solution is derived from the best solutions stored at the fanin nodes of the matching gate. This is true for power estimation under a zero delay model, but not for that under a general delay model. Consider figure 7.9 as an example. Let M_1 and M_2 be two mapping candidates at n with the same delay. Let n have a larger $E_n(switching)$ value for M_1. It appears that M_2 is superior to M_1 and thus M_1 is dropped form the power-delay curve.

However, when doing mapping at m, the mapped-parent of n, due to the difference in the transition waveform timing for M_1 and M_2 with respect to the other inputs of m, the best solution at m may be coming from M_1 instead of M_2 (7.9 b). So if M_1 is dropped from the power-delay curve, the optimal solution may not be found. The reason is that glitches depend on the gate delay and the transition waveforms of the gate inputs which are not related to the minimum power mapping solutions for the inputs. It is very expensive to store all partial solutions, so we have to drop the locally inferior solutions. The dynamic programming approach acts only as a heuristic approach (even for trees).

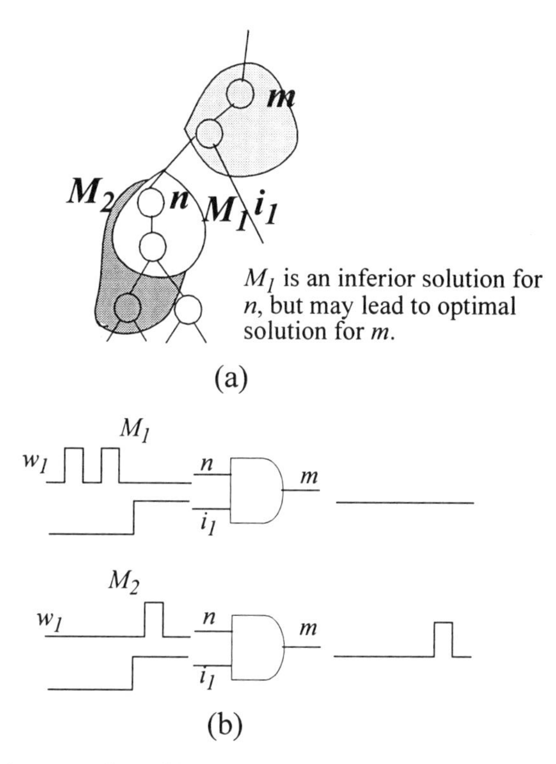

Figure 7.9 An example to illustrate complications due to a general delay model.

7.3.7 Extension to consider signal correlation at the primary inputs

One assumption we made was that primary inputs are pairwise uncorrelated. If this assumption is relaxed, we must use the conditional probabilities to calculate the signal probability at the outputs of intermediate nodes. It is, however, too costly to consider conditional probabilities among all subsets of primary inputs. Instead, we can only use pairwise conditional probabilities as follows. Correlation coefficient $C(i,j)$ is defined by

$$P(ij) = P(i)P(j|i) = P(i)P(j)C(i,j) \tag{7.17}$$

that is,

$$C(i,j) = \frac{P(i|j)}{P(i)} = \frac{P(j|i)}{P(j))} \quad (7.18)$$

Ercolani et al. [9] describe a method to approximate the correlation coefficients of the outputs of gates given the signal probabilities and correlation coefficients of the inputs as follows.

Let g be a gate with inputs i and j and the correlation coefficients of i and j, i and m, and j and m be given as $C(i,j)$, $C(i,m)$ and $C(j,m)$. The correlation coefficient of g and m is approximated by

g = AND gate:

$$C(g, m) = C(i, m)C(j, m) \quad (7.19)$$

g = OR gate:

$$C(g, m) = \frac{P(i)C(i, m) + P(j)C(j, m) - P(i)P(j)C(i, m)C(j, m)C(i, j)}{P(i) + P(j) - P(i)P(j)C(i, j)} \quad (7.20)$$

g = NOT gate:

$$C(g, m) = \frac{1 - P(i)C(i, m)}{1 - P(i)} \quad (7.21)$$

The signal probability of a product term is estimated by breaking down the implicant into a tree of 2-input AND gates and then using the above formula to calculate the correlation coefficients of the internal nodes and hence the signal probability at the output. Similarly, the signal probability of a sum term is estimated by breaking down the implicate into a tree of 2-input OR gates.

7.4 Power Reduction after Technology Mapping

Further power reduction can be obtained after technology mapping. Here we have a mapped circuit whose power consumption and delay can be modeled accurately. Two different approaches exist which modify a given mapped circuit to reduce the power consumption without increasing the delay. The first approach re-sizes the gates which have possible slack to reduce the power consumption while satisfying the delay constraints are kept [2]. The structure of the mapped circuit remains the same, only the size of some of the gates are changed. The second approach reduces power dissipation after technology mapping by structural optimization of the netlist. It performs a sequence of permissible signal substitutions, where each substitution reduces the power consumption of the circuit [21]. Structural optimizations are discussed in chapter 8.

7.4.1 Gate resizing

Power dissipation depends on the capacitance loading. Logic gates having smaller size lead to smaller power consumption. On the other hand smaller gates are slower. For a mapped circuit, not every gate is on the critical path. Gates that are not on the critical path, have a positive slack which is defined as the difference between the required time and actual arrival time at the output of a gate. The sizes of these gates can be reduced to minimize the power consumption without affecting the overall performance of the circuit.

In [1], a symbolic procedure based on Algebraic Decision Diagrams (ADDs) was proposed to accurately calculate the arrival time at the output of each gate for any primary input vector. In [2], the approach was extended to calculate the required times and slack symbolically. False paths can automatically be detected using this approach.

7.4.1.1 Timing calculation

Usually static timing analysis is used to provide timing information when we calculate the arrival and required time at the output of the gate. However, the value is not always accurate since it ignores the dependence on the input values and hence does not consider the false path. In fact, for circuits with false path, static timing analysis will give an overly pessimistic value. When considering gates for re-sizing, some of the gates may be obmitted since the slacks are over-pessimistic. This will limit the potential for further power reduction. In [1] and [2], Algebraic Decision Diagrams (ADD), which can be used to compactly represent functions from $\{0,1\}^n$ for large value of n, are used to store the arrival time at a gate's output for all input vectors. The arrival time at the output of a gate is calculated as follows. Given a gate g of the network and a primary input vector x, the arrival time at its output line $AT(g,x)$, is calculated based on the arrival time of its inputs and the delay of its fanin connections, $d(c_j,x)$ where c_j is the connection to pin j of gate g. If at least one fanin c_j of g has a controlling value of input x where x is a possible care input vector, then

$$AT(g,x) = min_j\{AT(c_j,x) + d(c_j,x)|(c_j = controlling)\} \tag{7.22}$$

If all fanins of g have non-controlling values, then

$$AT(g,x) = max_j\{AT(c_j,x) + d(c_j,x)\} \tag{7.23}$$

If x is not a possible care input vector, $AT(g,x) = -\infty$.

The required time can be calculated in a similar way. Let $E = \{e_j\}$ be the set of fanin connections for gate g and $C(x) \subseteq E$ be the subset of fanin connections that carry controlling values for a possible care input vector x. If all inputs to g are non-controlling, i.e., $C(x) = \emptyset$, then

$$RT(e_j,x) = RT(g,x) - d(e_j,x) \qquad \forall e_j \in E \tag{7.24}$$

If some inputs to g are controlling, let $c_d \in C(x)$ be a designated connection of which $A(c_d,x) \le RT(g,x) - d(c_d,x)$, then

$$RT(c_d,x) = RT(g,x) - d(c_d,x) \qquad and \tag{7.25}$$

$$RT(e_j,x) = \infty \qquad \forall e_j \in E, e_j \neq c_d \tag{7.26}$$

If x is not a possible care input vector, then $RT(e_j,x) = \infty$. Let $F = \{f_j\}$ be the set of all fanout connections for gate g, then

$$RT(g,x) = min\{RT(f_j,x)\} \qquad \forall f_j \in F \tag{7.27}$$

Once the arrival and required time calculation are finished, the slack of each gate g for an input vector x, $ST(g,x)$, can be computed as follows:

$$ST(c_j,x) = RT(c_j,x) - AT(g,x) \tag{7.28}$$

$$ST(g,x) = min_j\{RT(f_j,x)\} - AT(g,x) \qquad \forall f_j \in F \tag{7.29}$$

Different ADDs are built to store the $AT(g,x)$, $RT(g,x)$ *and* $ST(g,x)$ at the output of the gate g.

7.4.1.2 Gate re-sizing algorithm

The gate re-sizing algorithm is combined with the required time and slack calculation. The actual slack of a gate can be calculated by

$$slack(g)_{rise} = min_x\{ST(g, x)\}, \forall x \in \text{set of all possible care input vector}, f(g, x) = 1 \quad (7.30)$$

$$slack(g)_{fall} = min_x\{ST(g, x)\}, \forall x \in \text{set of all possible care input vector}, f(g, x) = 0 \quad (7.31)$$

If a gate has a positive slack, it can be re-sized with a smaller equivalent gate from the library to reduce the capacitance loading. Once the gate is resized, the required time at its output will not change. The required time and the slack of the gate's fanin gates are updated using the new gate delay information. The algorithm starts from the primary output and then works back to the primary inputs in a pre-order traversal.

7.4.1.3 Experimental results

Results of using gate-resizing to reduce the power consumption of mapped circuit were presented in [2]. The results are summarized in Table 7.4. The experiment results was obtained on some of the **MCNC**'91 benchmarks [18]. The circuits were optimized for area with the SIS *script.rugged* and mapped with the *SIS map -n 1 -AFG* command. The library used in this experiment had NANDs, NORs and inverters with inputs up to 4. For each type of gate, there existed 5 different sizes. The results of using static time analysis are also included for comparison with that using ADD. It is shown that using the gate re-sizing, the power reduction compared with minimum area mapped circuit ranges from 9% to 38%. The amount of power reduction is expected to be less if the circuit is originally mapped for low power.The power saving is also smaller if static timing analysis is used instead of ADD timing analysis.

Cir.	Gates	Init.Static Rise/ Fall	Init.ADD Rise/ Fall	Method	Final Static Delay Rise/Fall	Final ADD Delay Rise/Fall	Power Saved (%)
5xp1	174	15.88/	15.63/	ADD	17.15/17.15	15.69/15.69	11.54
		15.88	15.74	static	15.82/15.82	15.73/15.74	8.22
bw	241	17.58/	17.21/	ADD	20.92/20.92	17.18/16.66	31.89
		17.58	16.71	static	17.53/17.53	17.46/17.35	3.03
clip	192	12.29/	12.22/	ADD	13.45/13.45	12.22/12.20	15.68
		12.29	12.24	static	12.29/12.29	12.17/12.20	2.95
rd73	96	15.10/	14.85/	ADD	15.50/15.50	14.94/14.97	10.89
		15.10	15.05	static	15.10/15.10	15.00/15.05	3.75
sao2	213	15.46/	15.46/	ADD	16.58/16.58	15.43/15.41	13.38
		15.46	15.40	static	15.46/15.46	15.46/15.38	6.21
sct	132	16.96/	16.96/	ADD	20.99/20.99	16.88/16.86	9.52
		16.96	16.96	static	16.96/16.96	16.96/16.96	1.44
squar5	101	13.75/	13.46/	ADD	16.68/16.68	13.26/13.39	38.23
		13.75	12.61	static	14.01/14.01	13.72/13.02	22.03
ttt2	306	11.36/	11.36/	ADD	12.36/12.36	11.29/11.16	38.95
		11.36	11.16	static	11.28/11.28	11.28/11.28	37.61
cbp8	186	19.30/	15.28/	ADD	21.58/21.58	15.17/15.17	14.81
		19.30	15.28	static	19.25/19.25	15.61/15.61	0.51
mult4	202	24.41/	22.94/	ADD	26.19/26.19	22.89/21.24	23.78
		24.41	21.05	static	24.34/24.34	23.57/21.95	9.44

Table 7.4 Comparison of Power Savings by gate-resizing on Circuits with False Paths using ADD versus Static Timing Analysis [2]

Reference

[1] R. I. Bahar, E. A. Frohm, C. M. Gaona, G. D. Hachtel, E. Macii, A. Pardo and F. Somenzi, "Algebraic decision diagrams and their applications", in the proceedings of IEEE/ACM International Conference of Computer-Aided Design, pp. 188-191, November, 1993.

[2] R. I. Bahar, G. D. Hachtel, E. Macii and F. Somenzi, "A symbolic method to reduce power consumption of circuits containing false paths", in the proceedings of IEEE/ACM International Conference of Computer-Aided Design, pp. 368-371, November, 1994.

[3] R. Bryant, "Graph-based algorithms for Boolean function manipulation", in IEEE Transactions on Computer-Aided Design vol. C-35, pp. 677-691, August, 1986.

[4] R. K. Brayton and R. Rudell and A. Sangiovanni-Vincentelli and A. Wang, "MIS: A multiple-level logic optimization system" in IEEE Transactions on Computer-Aided Design, vol. CAD-6, No.6. pp. 1062-1081, November, 1987.

[5] R. K. Brayton and G. D. Hachtel and A. L. Sangiovanni-Vincentelli, "Multilevel logic synthesis", in Proceedings of IEEE, vol. 78, pp. 264-300, February, 1990.

[6] K. Chaudhary and M. Pedram, "A near-optimal algorithm for technology mapping minimizing area under delay constraints", in the proceedings of IEEE/ACM Design Automation Conference pp. 492-498, June, 1992.

[7] J. Darringer and D. Brand and J. Gerbi and W. Joyner and L. Trevillyan, "LSS: A system for production logic synthesis" IBM Journal of Research and Development, vol.28 no. 5 pp. 537-545, September, 1984.

[8] E. Detjens and G. Gannot and R. Rudell and A. Sangiovanni-Vincentelli and A. Wang, "Technology mapping in MIS", in the proceedings of IEEE/ACM International Conference of Computer-Aided Design, pp. 116-119, November, 1987,.

[9] S. Ercolani and M. Favalli and M. Damiani and P. Olivo and B. Ricco, "Estimate of signal probability in combinational logic networks", in the proceedings of European Testing Conference, pp. 132-138, 1989.

[10] A. A.Ghosh and S. Devadas and K. Keutzer and J. White, "Estimation of average switching activity in combinational and sequential circuits", in the proceedings of IEEE/ACM Design Automation Conference pp. 253-259, June 1992.

[11] D. Gregory and K. Bartlett and A. de Geus and G. Hachtel, "SOCRATES: A system for automatically synthesizing and optimizing combinational logic", in the proceedings of IEEE/ACM Design Automation Conference pp. 79-85, June, 1986.

[12] J. Hayes "A NAND model for fault diagnosis in combinational logic networks", in IEEE Transactions on Computer, vol. C-20, pp. 1496-1506, 1971.

[13] D. A. Huffman, "A method for the construction of minimum redundancy codes" in Proceedings of the IRE, vol. 40, pp. 1098-1101, September, 1952.

[14] A. Aho and S. Johnson, "Optimal code generation for expression trees", in Journal of ACM. pp. 488-501, July 1976.

[15] K. Keutzer, "Dagon: Technology binding and local optimization by DAG matching" in the proceedings of IEEE/ACM Design Automation Conference pp. 341-347, June, 1987.

[16] L. Larmore and D. S. Hirschberg, "A fast algorithm for optimal length-limited Huffman codes", Journal of ACM, Vol. 37, No. 3, pp. 464-473, March, 1990.

[17] J. Ishikawa and H. Sato and M. Hiramine and K. Ishida and S. Oguri and Y. Kazuma and S. Murai, "A rule based reorganization system LORES/EX" in the proceedings of IEEE/ACM International Conference on Computer Design, pp. 262-266, October, 1988.

[18] S. Yang, "Logic synthesis and optimization benchmarks user guide version 3.0," Technical Report, Microelectronics Center of North Carolina, Research Triangle Park, NC, January, 1991

[19] F. Najm "Transition density: A stochastic measure of activity in digital circuits", in the proceedings of IEEE/ACM Design Automation Conference pp. 644-649, June 1991.

[20] D.S.Parker Jr., "Conditions for optimality of the Huffman algorithm" "SIAM Journal of Computing, vol 9, no. 3, pp. 470-489, August, 1980.

[21] B. Rohfleisch, A. Kolbl and B. Wurth, "Reducing power dissipation after technology mapping by structural transformation", in the proceedings of IEEE/ACM Design Automation Conference, pp. 789-794, June, 1996.

[22] R. Rudell "Logic synthesis for VLSI design", University of California, Berkeley Ph.D. Thesis, 1989.

[23] H.Savoj and H.-Y.Wang, "Improved scripts in MIS-II for logic minimization of combinational circuits", in the proceedings of the International Workshop on Logic Synthesis", May, 1991.

[24] A. A. Shen and A. Ghosh and S. Devadas and K. Keutzer, "On average power dissipation and random pattern testability of CMOS combinational logic networks", in the proceedings of IEEE/ACM International Conference of Computer-Aided Design, pp. 402-407, November, 1992.

[25] K. J. Singh and A. Wang and R. K. Brayton and A. Sangiovanni-Vincentelli, "Timing optimization of combinational logic", in the proceedings of IEEE/ACM International Conference of Computer-Aided Design, pp. 282-285, Nov, 1988.

[26] L. Stockmeyer, "Optimal orientations of cells in slicing floorplan designs" Information and Control, vol 57, pp. 91-101, 1983.

[27] H.J.Touati and C.W.Moon and R.K.Brayton and A.Wang, "Performance-oriented technology mapping", in the proceedings of the Sixth M.I.T. Conference on Advanced Research in VLSI, pp. 79-97, April, 1990.

[28] C. Y. Tsui and M. Pedram and A. Despain "Efficient Estimation of Dynamic Power Consumption under a Real Delay Model", in the proceedings of IEEE/ACM International Conference of Computer-Aided Design, pp. 224-228, November, 1993.

CHAPTER 8

Post Mapping Structural Optimization for Low Power

Traditionally, logic synthesis has been divided into two stages: technology independent optimization and technology mapping. Technology independent power optimization techniques have been presented in the previous chapters of this book. The structural optimization techniques have recently been introduced as a new optimization step. In a multi-level network, it is, in general, possible to replace a signal (a node input or output) with an existing signal in the network. At the same time, it is also possible to replace a signal with a combination of two or more signals in the networks. In doing so, it is possible to optimize different costs in the network. For example, delay optimization is performed by replacing a signal in a network with an equivalent signal when this replacement will result in reducing the critical path in the network. In [2], a method is presented where a Boolean network is optimized by first inserting redundancies in the network and then removing other sets of redundancies which lead to a lower total cost of the network. In [1], a method is presented that attempts to identify alternative wires and alternative functions for a target wire that cannot be routed due to the limited routing resources in an FPGA. Alternative wires that can be routed through less congested areas substitute the un-routable wires without changing the circuit's functionality. Redundancy addition and removal techniques have again been used to find alternative wires in the circuit. In both these methods ATPG techniques have been used to check for functional correctness of the circuit when checking for candidate alternative wires. The techniques presented in [1], [2] and [6], have been extended in [7] to minimize the power consumption of Boolean network. The techniques for structural optimization have been applied after the Boolean network is mapped to the target technology. The reason for this is that cost functions can more accurately be calculated for a netlist of gates since all technology dependent information such as area, delay, and load values are readily available from the netlist of gates.

This chapter presents techniques for minimizing the power consumption of netlist of gates based on structural transformations. Techniques presented in this chapter are based on an incremental approach where observability don't cares in the network are first calculated and then possible substitutions are explored and used when power consumption can be reduced by making the substitution.

8.1 Signal Substitution

A *wire* in a netlist of gates corresponds to an edge in the graph representing the Boolean network corresponding to the netlist of gates. *Source* of a wire corresponds to the gate that drives the wire and *sink* of a wire corresponds to the gate that the wire in driving. Replacing a wire w_1 with a wire w_2, in the most general case, refers to adding a new wire w_2 to the network where this addition allows wire w_1 to be removed. *Source-replacing* a wire w_1 with a wire w_2 refers to replacing the sink of wire w_1 with the source for wire w_2. Note that source-replacing wire w_1 with a wire w_2 is a special case of wire replacement where the sink for wire w_2 is constrained to be the sink for wire w_1. Source-replacing is also defined as input-substitution. *Output substitution* of gate g with wire w refers to source-replacing all fanout wires of g with wire w. A *target wire* is defined as a wire that is considered for removal. An *alternative wire* is a wire that is used to replace the target wire.

Consider the example in figure 8.1. As can be seen in this example, function f can be expressed both in terms of inputs a and d and also inputs e and d. This flexibility in expressing the function of f can be exploited to optimize different costs in the network. For example, if path *a-e-f* in configuration A is on the critical path, then delay can be reduced by replacing wire e with wire a and therefore reducing the circuit delay. If the switching activity of wire a is less than the switching activity of wire e, then power can also be reduced by replacing wire e with wire a.

The main task in performing structural optimization is to find the set of possible wire replacements for a given wire in the network. Once the set of candidate replacement wires has been identified, the best replacement can be selected by assigning a value to each candidate wire which gives the contribution of this replacement to the cost of the network being minimized.

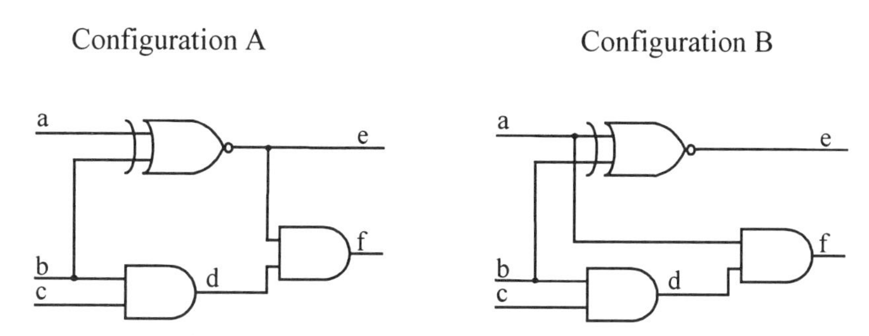

Figure 8.1 Replacing a wire with another wire in the network.

A number of approaches have been presented in the past for performing structural optimization. The approach presented in [1] uses redundancy addition/removal techniques to find alternative wires for a given wire in the network. Based on this approach, the alternative wire, when added to the network, makes the target wire redundant. A wire is redundant if and only if the corresponding stock-at fault is un-testable. The concept of mandatory assignments introduced in [3] is used in [1] to find alternative wires for a target wire. When testing a wire for a stuck-at fault, the set of mandatory assignments are the value assignments required for a test vector to exist for this stuck-at fault. The set of mandatory assignments are computed in [1] by using implication [4], [5], and recursive learning [8] techniques. If the set of mandatory assignments for a wire cannot be satisfied, then the corresponding stuck-at fault is redundant. Therefore, a target wire is made redundant by adding wires to the network that will make the set of mandatory assignments for this wire inconsistent. It should be guaranteed that adding the new wire will not change the behavior of the circuit. This can be guaranteed by checking that adding the new connection forms an un-testable stuck-at fault. Therefore, each new wire is checked to be a redundant wire before the target wire is removed.

The approach used in [7] is based on logic clause analysis [6]. A clause is defined as sum of variables and a valid clause describes dependencies among its variables. A clause is said to be valid if, and only if, it evaluates to 1 for all assignments to the primary inputs of the circuits. Global implications introduced in [8] correspond to valid global clauses which cannot be derived from a single gate's formula. A valid global clause describes global signal dependencies and, therefore, can be used to optimize the cost that is being minimized. In [7], bit parallel fault simulation (BPFS) [9] and automatic test pattern generation (ATPG) techniques have been used to compute a set of valid clauses which correspond to single wire or a combination of wires that can be used to replace a target wire. This information is then used to minimize the delay in the network.

Previous techniques for structural optimization have been based on test pattern generation and fault simulation techniques. The next section presents a technique for performing post-mapping structural optimization by using the observability don't care conditions which are derived from the network structure.

8.2 Candidate Wire Generation

Redundancy addition followed by redundancy removal is the underlying technique in structural network optimization. In general, it is possible to add redundancy in one part of the network followed by removing redundancies from a different part of the network. The approach presented in [1] makes use of sets of mandatory assignments to find the redundancy that needs to be added to a network in order to remove a target wire. The time complexity of this operation, however, proves to be very high. A different approach for removing a wire *w* is to only consider the set of wires that can be used for source-replacing wire *w*. This technique is equivalent to redundancy addition/removal where the added redundancy is limited to the same gate that the target wire is used in. Using this restriction, the problem is formulated as follows:

Problem: *Given a Boolean network mapped to the target technology and a target wire w (or a target gate g), find all possible wires that can be used to input substitute wire w (or output substitute gate g).*

The goal in structural optimization is to take advantage of the redundancy present in the current implementation of the network being optimized. Therefore, the first step in performing structural optimization is to consider substitutions that involve only a single wire. An extension of this technique is to also consider wires that are combinations of other existing wires in the network. Experimental results show that considering existing wires does not provide enough flexibility in optimizing the cost of the network. In order to provide more wires that may be considered as replacement wires, complement of existing wires are also considered. In addition, a new set of wires are generated by combining pairs of existing wires in the network. This operation is generalized as follows: For every pair of wires x and y, all possible functions of these two variables (which include x, y and their complement and excluding constant 0 and 1) are also considered as candidate substitutions.

8.3 Cost functions

The goal of structural optimization is to reduce the switched capacitance of a Boolean network after it is mapped to the target technology. Since the network is already mapped to the target technology, it is possible to find an accurate cost for each possible wire replacement operation. In this section, a cost model is presented for taking into account all factors affecting power consumption when a wire is replaced with another wire in the network.

Replacing a wire in a netlist of gates will result in changing the total switched capacitance of the circuit. Figure 8.2 shows an example where output of gate g_1 is substituted with wire d which is the output of a new gate formed by forming the AND of wires b and c. Three sources of change in power can be identified in this example: 1) power reduction due to removing gate g_1 and all nodes in its exclusive transitive fanin cone, 2)power increase due to inserting gate g_2 in the network which results in increased load on its inputs and 3) the power change due to the change in the switching activity values in the transitive fanout cone of gate g_2. This example shows the change in power for an output substitution. Note that because of the change in the load values on nodes b, c and also nodes that drive nodes in region A, the delay values in the network will change and this will in turn change the power due to spurious logic value changes. This change in power can be accounted by measuring the change in the hazardous behavior of the network.

The value of a wire substitution gives the reduction in the power consumption of the network when the substitution is performed and is computed using equation 8.1

$$Value = \sum_{w \in A} sw(w) \cdot C_w - \sum_{w \in B} sw(w) \cdot C_w + \sum_{w \in C} \Delta sw(w) \cdot C_w \tag{8.1}$$

where regions A, B and C are shown in figure 8.2 and correspond to the three sources of change in power. *sw(w)* corresponds to the switching activity of wire w and C_w gives the load seen by wire **w**. Change in switching activity in the last part of this equation corresponds to the change in the switching activity of wires in the transitive fanout of gate g_1 as it is replaced with the output of gate g_2. Note that input substitution is a special case of output substitution and the value for an input substitution can be computed using the same equation except that region A will be an empty region. The reason is that for an input substitution, only one of the fanout edges of a gate is replaced with another wire and, therefore, the gate is required to drive the other fanout edges. Therefore no gate can be removed from the circuit after an input substitution

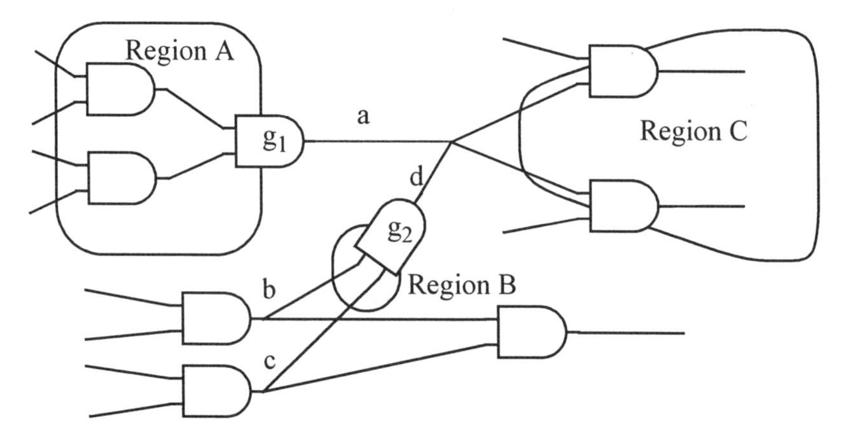

Figure 8.2 Change in power consumption after a substitution

unless the gate has only one fanout in which case input substitution and output substitution will be equivalent operations. At the same time, for a single wire substitution (without needing an inverter), region B will be an empty region since no new gates will be added to the circuit after the substitution.

8.4 Function Substitutions

In a Boolean network, in order to replace a wire w with another wire z, it is necessary that, after replacing wire w with wire z, the behavior of the circuit at the outputs is not changed. As pointed out before, the problem of structural optimization is simplified by only looking at wire substitutions that replace the target wire at the sink. This means that the function of the sink gate for the target wire need not be modified. In this context, the problem of identifying a replacement wire for a target wire is formulated as follows: Given a target wire w_1 and a wire w_2, determine if w_2 can replace w_1.

In the absence of observability don't care conditions, this substitution is only possible if the global function of w_2 is the same as the global function of w_1. In this case, wire w_1 can be replaced with wire w_2. Even though this condition is easy to check using BDDs, in general the number of such wires is very small. The conditions for replacing wires can be relaxed by taking into account the observability don't care conditions for wire w_1. Given F_1, the global function and ODC, the observability don't care conditions for wire w_1 and F_2, the global function of wire w_2, then wire w_1 can be replaced with wire w_2 if and only if:

$$F_1 - ODC \subseteq F_2 \subseteq F_1 + ODC \qquad (8.2)$$

Once the ODC is calculated, this condition is easily checked for by using global BDDs. Once a possible candidate substitution is identified, the value for this substitution is computed using equation 8.1 and compared to other possible substitutions.

A number of issues need to be considered while searching for candidate substitutions. These issues include the techniques for computing ODCs, increasing the search space for finding candidate substitutions, and computational complexity of the optimization process. These issues are discussed in the following sections.

8.4.1 Computing ODCs

Computing ODCs for low power has been discussed in detail in previous chapters. By using ODCs computed for low power, it is guaranteed that changes in the function of a node will not result in increasing the power consumption in the transitive fanout of the node. This means that it is not necessary to analyze the effect of changes in the switching activity of transitive fanout nodes. Using ODCs computed for low power, it is also guaranteed that any wire substitution will not increase the power consumption in the transitive fanout nodes. Therefore, it is possible to only consider the effect of power changes due to regions A and B in figure 8.2. It should however, be noted that even though all possible substitutions will never result in an increase in the power consumption of nodes in the transitive fanout nodes, some substitutions may reduce the power in the transitive fanout nodes more than other substitutions. Therefore, it is worthwhile to also consider this component of power reduction in selecting one substitution over another one. At the same time it should be noted that using ODCs computed for low power will, generally, reduce the number of possible substitutions which may potentially lead to a large decrease in power by removing some gates from the network which fall into region A in figure 8.2. Therefore, if a decision is made to also take into account the change in power in the transitive fanout of a node, then ODCs can be computed without regard for power consumption. However, if it is too expensive to keep track of this change in the transitive fanout power, then ODCs should be calculated for low power.

ODC computation procedures should also be extended to be used with input substitution. ODC computation techniques discussed previously can directly be used for output substitution since the ODC computed for the gate that is driving the target wire gives the ODC conditions for the target wire. The size of ODC for a multiple fanout gate is usually very small. This is because the ODC for such a gate is computed by finding the intersection of ODC conditions for each fanout edge. This means that even though it may not be possible to replace the output of a given gate in the circuit, it may be possible to perform input substitution on some of its fanout edges. Since the ODC for a gate is computed by first computing the ODC for each of its fanout edges, then it is straight forward to use the ODC for each fanout edge (or each fanout wire) to find a candidate substitute for that target wire. Since an output substitution, in general contributes more to reducing the circuit power (by removing some nodes in the circuit) during the procedure for each gate we first check for possible output substitutions and if this operation fails, input substitution for each fanout edge of the gate is considered.

8.4.2 Increasing number of candidate substitute wires

As pointed out earlier, only considering the existing wires in the circuit as candidate substitute wires will not provide enough flexibility in performing structural optimizations. A substitute wire can usually be generated by combining any number of existing wires in the circuit using any function of these wires. This will, however, generate a large number of candidate wires to be checked as substitute wires. Given a target wire w and m other wires that may be considered as substitute wires, there are 2^{2^m} possible functions to check for substitute wires. This number is very large and cannot, in practice, be considered. It should also be noted that the structural optimization is being performed on a technology mapped network and, therefore, adding any function to the network will require mapping of the new part and computing the associated cost. Experimental results show that considering all functions of pairs of wires will in general provide reasonable amount of flexibility for finding substitute wires. In order to obtain a good balance between flexibility and efficiency we choose to only look at all functions of pairs of wires in the circuit. This means that for every pair of wires x and y, all possible functions of these two variables (which include x, y and their complement and excluding constants 0 and 1) are checked as candidate substitutions.

When substituting a single wire for another wire, it is not necessary to insert a new gate in the circuit since the modification consists only removing one wire and adding a new connection in its place. On the other hand, when a combination of two wires is used to replace another wire, it is necessary to insert new gates in the circuit. For example, if a library does not contain AND gates, then replacing a wire with the AND of two wires requires that a NAND gate and an inverter to be inserted in the circuit. The addition of these gates corresponds to the power increase in region B in figure 8.2. Technology libraries usually contain gates that are variations of simple NAND and NOR gates. For example, a library may also contain a AOI gate with equivalent function $\overline{a \cdot c + b}$. Therefore, if an input to a NOR gate needs to be replaced with the AND of two other wires in the circuit, it is possible to change the NOR gate to an AOI gate. Using local re-mapping techniques such as the one illustrated in this example it is possible to further minimize the power cost of the circuit.

8.4.3 Functional simulation to speed up optimization

The problem with the approach presented so far is that it is very expensive to perform functional equivalency as proposed in section 8.4. This is because operations on global BDDs are very expensive for large circuits and, therefore, it is not possible to check a large number of candidate wires as possible substitute wires.

Functional simulation can be used to significantly speedup the functional equivalence check as described next. In order to speed up the algorithm, a bit-parallel functional simulation package based on gate level description of the circuit is developed. A similar package has also been developed for functional simulation using BDDs. Functional simulation can be performed at very high rates. For example, 5-10 million gate-vector/second can be simulated on a sparc20 workstation using the gate level functional simulation package. The optimization procedure is augmented as follows. First, a set of random vectors are applied at the circuit primary inputs and then the value of each gate in the circuit is computed for each of these vectors. For each candidate wire to be considered as a possible substitute wire, the equivalence of the corresponding simulation vectors for the replacing wire and the target wire is checked using equation 8.2. For this comparison, the simulation vector for ODC is also obtained using bit-parallel functional simulation. If the result of this check implies that the candidate wire cannot be substituted for the target wire, it is guaranteed that this substitution is not possible. On the other hand if the check is positive, the global functions of wires need to be checked to guarantee that for all minterms in the space of the circuit primary inputs the wires are functionally equivalent within the ODC of the target wire.

Experimental results show that the number of global BDD compare operations is significantly reduced for a small number of vectors simulated at the input. For example, for 256 vectors simulated at the circuit inputs, the number of required checks using global BDDs is reduced by two orders of magnitude.

Filters based on the support of a function can also be used to limit the number of checks that need to be made. Experimental results show that in majority of cases, if a candidate wire can be used as a substitute wire, the support of this candidate wire in terms of the primary inputs of the circuit contains the primary input support of the target wire that is being replaced. Therefore set of wires whose support does contain the support of the target wire can be eliminated from the set of wires to be checked as substitute wires.

8.5 Experimental results

The procedures presented in this chapter were implemented and applied to example benchmarks after they were mapped to the target library. Each example in the benchmark set is first optimized for power and then mapped for minimum power. Structural optimization was then performed on the mapped circuits.

Table 8.1 shows the results of the optimization after applying structural optimization on the mapped networks starting from multi-level MCNC benchmarks. Columns 2 and 3 give the area and power of the network right after the circuits are mapped. The area is given as the total active area of the gates in the circuit and the power is estimated using the library load parameters. Columns 4 and 5 give the network area and power after structural optimization (normalized with respect to the values in columns 2 and 3) where area and power are again computed using the library parameters. Power is measured under a zero delay model and randomly set input signal probabilities. Results show that power is on average reduced by %11, while the area is reduced by %9. Table 8.2 provides the same results starting from two-level MCNC benchmarks. The results show that on average power is reduced by %14, while area is reduced by %4.

Examples	Before optimization		After optimization	
	Area	Power	Area	Power
C5315	1721904	56.61	0.92	0.88
C880	405072	10.82	0.97	0.98
comp	172144	4.55	0.85	0.81
des	4044224	89.77	1.01	0.93
example2	391616	7.13	0.95	0.95
pair	1858320	42.71	0.96	0.95
term1	169360	3.84	0.86	0.92
vda	798544	6.03	0.96	0.93
9symml	264944	5.37	0.99	0.96
C1355	434304	14.89	0.86	0.84
C1908	542416	12.92	0.81	0.85
C432	257520	7.10	0.83	0.79
alu2	502048	8.01	0.91	0.83
alu4	996208	7.15	0.97	0.90
cordic	83056	2.03	0.83	0.86
dalu	1335856	26.55	0.87	0.88
f51m	163792	3.45	0.84	0.86
frg2	1063488	15.91	0.89	0.87
k2	1482016	8.14	0.95	0.86
t481	771632	5.07	0.95	0.93
Average			0.91	.89

Table 8.1 Area, delay and power for post mapping optimization (multi-level examples)

Examples	Before optimization		After optimization	
	Area	Power	Area	Power
apex1	1762736	22.04	0.98	0.83
apex2	409248	6.15	0.96	0.92
cps	1396640	13.91	1.00	0.87
misex3	849584	11.42	0.99	0.86
pdc	676512	13.04	0.94	0.86
spla	767920	10.42	0.95	0.83
table3	1042608	13.94	0.99	0.86
table5	1235168	11.92	1.01	0.83
5xp1	109968	2.76	0.94	0.89
Z5xp1	132704	3.75	0.88	0.84
apex5	929856	9.29	0.97	0.91
bw	205552	2.93	1.00	0.82
clip	159152	4.44	0.95	0.94
duke2	472816	6.42	0.96	0.89
e64	294640	2.64	0.96	0.55
ex5	379552	9.24	0.93	0.88
rd53	36656	1.81	0.86	0.88
rd84	191168	3.19	0.98	0.86
sao2	173072	3.63	1.00	0.90
squar5	65888	0.88	1.02	0.93
Average			0.96	.86

Table 8.2 Area, delay and power for post mapping optimization (two-level examples)

References

[1] S. C. Chang, K. T. Cheng, N. S. Woo and M. Marek-Sadowska, "Layout driven logic synthesis for FPGAs." In proceedings of the *ACM/IEEE Design Automation Conference*, pages 308-313, 1994.

[2] K.T. Cheng and L. A. Entrena, "Multi-level logic optimization by redundancy addition and removal." In *European Conference on Design Automation*, pages 373-377, 1993.

[3] L. A. Entrena and K.T. Cheng. "Sequential Logic Optimization by redundancy addition and removal." In proceedings of the *IEEE International Conference on Computer-Aided Design*, pages 310-315, NOV 1993.

[4] T. Kirkland, M.R. Mercer. "A topological search algorithm for ATPG." In proceedings of the *ACM/IEEE Design Automation Conference*, pages 502-508, June 1987.

[5] W. Kunz and D. K. Pradhan. "Recursive learning: An attractive alternative to the decision tree for test generation digital circuits." In proceedings of the*International Test Conference*, pp816-825, October 1992.

[6] B. Rohfleisch, B. Wurth and K. Antreich. "Logic clause analysis for delay optimization." in proceedings of the *ACM/IEEE Design Automation Conference*, pages 668-672, June 1995.

[7] B. Rohleisch, A. Kolbl and B. Wurth. "Reducing power dissipation after technology mapping by structural transformations." In proceedings of the *ACM/IEEE Design Automation Conference*, June 1996.

[8] M. Schulz and E. Auth. "Advanced Automatic test pattern generation and redundancy identification techniques." In proceedings of the*Fault Tolerant Computing Symposium*, pages 30-34, June 1988.

[9] j. A. Waicukauski, E.B. Eichelberger, D.O. Forlenza, E. Lindbloom and T. McCarthy, "Fault simulation for structures VLSI." *VLSI Systems Design*, pages 20-32, 1985.

Part IV

Power Optimization Methodology

CHAPTER 9

POSE: Power Optimization and Synthesis Environment

In the past, the main objective for circuit designers has been to design fast and compact digital systems. In response to this demand, design automation tools have been developed to help the designers in automatic synthesis of digital circuits. In addition to synthesis tools, other automated programs have also been developed for simulation and verification of these synthesized circuits. By taking advantage of these tools and combining their features, designers have been able to develop and apply novel design methodologies enabling them to design faster and more complex systems while speeding up the design process. These design automation tools have been used extensively in the industry and are an integral part of any design cycle.

As described in chapter 1, the demand for low power digital systems has motivated significant research in the area of power estimation and power optimization. Power estimation and optimization techniques have been proposed at all stages of the design process. Even though considerable effort has been made in creating new techniques for power estimation and optimization, a unified framework for designing low power digital systems has not yet been developed. The void created by the absence of such a tool has presented designers with serious problems. Optimization algorithms that target low power circuits use the frameworks designed for synthesizing minimum area and delay circuits. This means that critical information needed for power estimation and optimization is not available when low power techniques are applied. In most cases, minimal information is made available to the power minimizing procedures which in turns results in reducing the capabilities of these design steps. The more significant problem faced by the lack of a unified framework for low power design is that designers are forced to use low power techniques as isolated procedures. This has been a major obstacle in developing a methodology for effective and efficient power specification, estimation and optimization. The lack of a methodology in turn, results in a limited understanding of the applicability of existing techniques which contributes to holding back the state of the art technology. At the same time, many optimization techniques are only applicable and relevant when applied in conjunction with other optimization approaches. This means that without a unified framework, many new techniques will not be discovered.

In order to address this problem, a complete system for designing low power digital circuit needs to be developed. The overall design flow for this process is shown in figure 1.2 where the initial design specification is optimized using the techniques provided at the behavioral, RT, logic and physical levels. The inputs to each level is the set of power relevant information which includes the necessary library information. The results of each stage is also checked using power prediction tools provided in the optimization environment.

We address this challenge by presenting a methodology for designing low power digital circuits at the logic level. In doing so, we present *POSE*, the *Power Optimization and Synthesis Environment*. POSE is the first step in creating a complete and unified framework for design and analysis of low power digital circuits. POSE provides a unified framework for specifying and maintaining power relevant circuit information. POSE also provides means of estimating power consumption of a circuit using different load models. POSE gives a set of options for making speed-accuracy trade-offs for estimating the power dissipation in a circuit and for making area-power trade-offs during circuit optimization. Example power optimization procedures include common-expression extraction, factorization, function decomposition and technology mapping. POSE provides an easy to use interactive environment which is an extension of the familiar environment provided by the SIS package [1].

This chapter presents an overview of the POSE system. Section 9.1 describes the low power design methodology used in POSE. Section 9.2 and 9.3 discuss issues behind design specification for power and power estimation. Section 9.4 presents an example of the power estimation flow used in POSE. Experimental results are presented in section 9.5. The final appendixes describe the file formats used in the POSE package.

9.1 Low Power Design Methodology

A low power design methodology can only be developed when a number of key components are made available to the designer. These components fall into the following categories: 1) *Design Specification*, 2) *Design verification* and 3) *Synthesis Procedures*. Complete design specification is necessary in order to provide the synthesis environment with maximum information necessary for the optimization process. For example, when designing a circuit for minimum area, the design specification for this synthesis process should include a measure of the area for the gates in the target library. Similarly, a synthesis environment for low power should include all the power related information that is necessary during the synthesis and validation procedure. An important issue to also consider is that design specification for power may not be the same at different levels of abstraction and therefore a set of consistent specification standards need to be available at different levels.

Design validation is also an important part of any design methodology. The final design has to comply with the design specifications. For conventional logic synthesis, this validation is in the form of checking functional correctness and checking that design meets the area and speed requirements. For power consideration, a design has to meet the target power consumption. This means that power estimation procedures are necessary to check design compliance with the given specifications. At the same time, power estimation is necessary for checking the quality of the synthesis steps. Power estimation is also a crucial part of interactive optimization techniques.

Synthesis procedures are the basic steps used to incrementally change the circuit structure while optimizing the target cost function. The combination of these steps guided by the power estimation procedures are used to develop a power optimization methodology.

Figure 9.1 presents the power optimization methodology used in POSE. The highlighted boxes specify the information that has to be provided by the designer. The input to POSE is a Boolean network specification describing the circuit to be designed. The set of power information required by design specifications (see section 9.2) is also provided at this time. The Boolean network is then optimized using a set of logic synthesis operations.

Low power optimization algorithms provided in POSE consist of three categories of optimization techniques: 1) Low power algebraic restructuring techniques (chapter 5), 2) Low power node simplification using power compatible don't cares (chapter 6), 3) Low power technology mapping (chapter 7) and 4) Post mapping structural optimi-

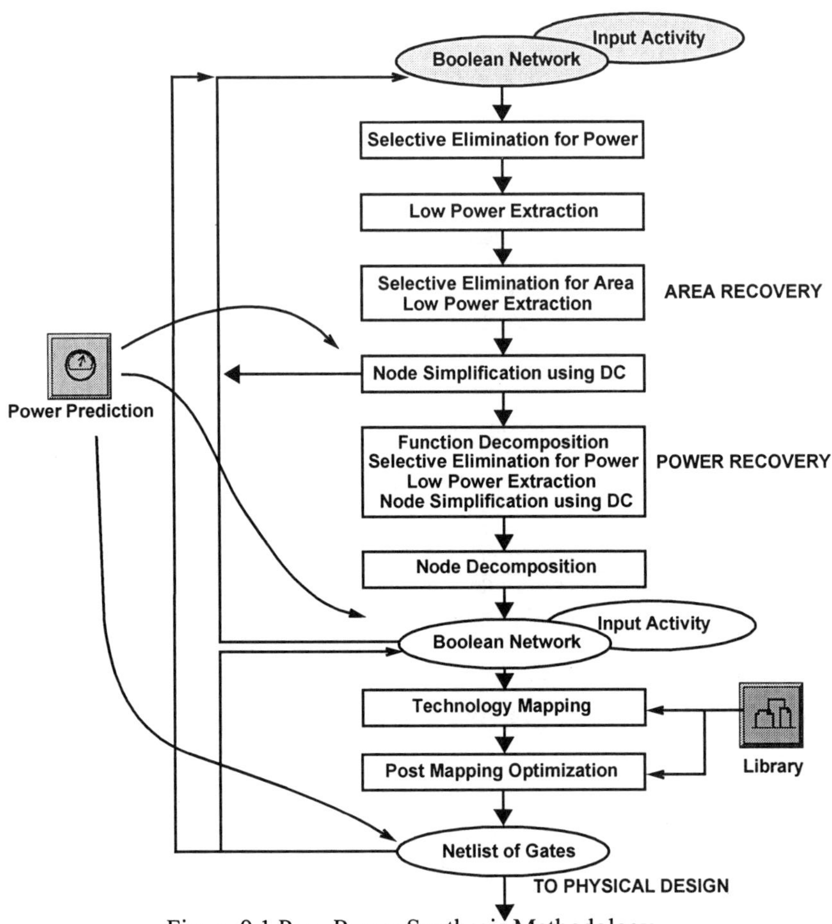

Figure 9.1 Pose Power Synthesis Methodology

zation techniques (chapter 8). Techniques in categories 1 and 2 are used during the logic synthesis process while the technology mapping is used to map the optimized networks to gates in the target technology. Post mapping structural optimization is applied to the network after it is mapped to the target technology.

Algebraic logic restructuring techniques have been used in the past to minimize the area of a Boolean network by taking advantage of the common sub-functions between different nodes and within the same node function. These operations include common sub-function extraction, function decomposition, and factorization. A selective collapse procedure is used to remove nodes from the network which don't contribute to minimizing the cost of the network.

POSE includes operations to perform low power logic restructuring. The low power algebraic operations implemented in POSE are discussed in chapter 5 and also in [4]. The basic approach here is to use extraction, factorization, decomposition, and selective collapse operations to create nodes that will maximally reduce the power in the network and also remove nodes that are not good candidate nodes for reducing the network power consumption.

It should be noted that power optimization techniques are developed to make area-power trade-offs. This is mainly done by reducing the power density[1] of the network so that even with the larger area, the total power consumption is smaller than the area optimized network. In fact, if a power optimization technique reduces power by reducing area, then it should be considered an area optimization technique. In the methodology given in figure 9.1, the initial logic restructuring has been performed for minimum power. This means that network area may not have maximally decreased during this operation. Therefore, the methodology includes a number of operations for recovering some area. This stage is called "*area recovery*". A "*power recovery*" stage is also included before the technology mapping process. The reason for this is that decomposition for minimum power does not maximally decompose all node functions. In other words it leaves some area redundancy in the network that may still be taken advantage of for power optimization. Therefore, the power recovery stage is included to use this flexibility.

An important issue to consider during power optimization is that the reduction in the power density of the network should more than offset the increase in the area of the network. Otherwise, this operation will result in a network with larger area and more power consumption. Therefore, it is desirable to be able to predict the potential effect of a power optimization step.

Among the logic restructuring operations discussed in this section, common sub-function extraction will, in general, result in most reduction in power with other operations having less impact on the final result. In the following we discuss how common sub-function extraction for low power may impact the final results.

---

[1]. Power density is defined as the total power divided by the area of the network.

Experimental results show that generally, if there exists more flexibility in extracting different subfunctions during the extraction procedure, the more power reduction can be obtained. This flexibility in extracting different subfunctions can be formalized in terms of the number of available subfunctions (kernels [6]) in the network before the extraction procedure is started. If this number is below a threshold value relative to the size of the network, then it may be concluded that a power optimization script using the extraction operation as its primary optimization step will not perform well and the regular minimum area script should be used.

After the Boolean network is optimized for power, it is then mapped to the gates in the target technology. This operation will generate a netlist of gates in the library. The goal of low power technology mapping is to generate a mapped network where high activity nodes are hidden inside complex gates. The procedures for low power technology mapping have been discussed in detail in [8] and chapter 7. After technology mapping is performed, it is possible to swap the symmetric inputs to a gate. The power after technology mapping can be further reduced by swapping pins where inputs with high input load are driven by lower activity inputs. By taking advantage of this flexibility, power can further be reduced. Technology mapping step is then followed by post-mapping structural optimization techniques as described in chapter 8.

Note that at some points during the optimization, the quality of the results are checked using the power estimation utilities. This is to check how much reduction in power has been obtained after each iteration. The synthesis process is terminated if no further improvement can be obtained, or the resulting power estimate is within the design specifications. The next section presents a detailed discussion on the design specification requirement and power estimation.

9.2 Design Specification for Power

In the past, logic synthesis has concentrated on minimizing the area of a circuit while meeting the timing constraints. The design specification for logic synthesis, therefore, consisted mainly of providing the functional description of the circuit, the timing constraints, and the area/delay characteristics of the target library. As logic synthesis environments are extended to take into account power consumption, the conventional design specification techniques prove to be inadequate. Information that is necessary for effective power estimation and optimization at the logic synthesis stage is presented in this section.

9.2.1 Input Switching Activity

The power consumption of a CMOS circuit is a function of the expected number of times the logic values in the circuit change values. This means that unlike conventional logic synthesis where the circuit performance (delay) is only affected by the *physical* characteristics of the surrounding logic (input drives and output loads), the transient behavior of the circuit has a significant impact on the power consumption of the circuit. In order to exactly capture the transient behavior of a circuit, it is necessary to provide a sequence of bit vectors applied at the primary inputs of the network. This sequence of bit-vectors may in turn be used by *statistical power estimation techniques* to compute an estimate of the expected switching rate of gates in a network, thus providing the necessary information for power estimation. Recently, probabilistic techniques have provided an efficient approach for capturing this transient behavior of circuits. By providing the expected behavior of the circuit at the primary inputs in terms of probability values, *probabilistic power estimation techniques [5]* provide an estimate of expected switching rates of the internal gates in a circuit.

The *swfile* format (Appendix A) is used to specify the primary input signal probability values.

9.2.2 Library Load Values

Libraries used during logic synthesis only provide information on area and delay of each gate in the library. More information is, however, required to accurately measure the power consumption of a gate in a technology mapped network. The power consumption of a gate consists of the power consumed at the output of the gate and the power internal to the gate.

The power at the output of a gate g is a function of the load seen at the output of this gate. This load is a combination of the input loads for output gates and also the *self-loading capacitance* for the gate itself. The self loading capacitance for a gate is defined as the load driven by the gate when the gate output is left open and is due to the source/drain diffusion capacitances of the gate. Experimental results show that self-loading capacitances contribute up to 20% of the total power consumption in CMOS circuits. Ignoring these capacitances will no doubt affect the accuracy of power estimation and the optimality of power optimization. The *internal power consumption* of a gate is computed by measuring the power required to charge and discharge the internal capacitances of this gate. This means that a power optimization procedure requires knowledge of the internal capacitances (diffusion capacitances) of all gates in the library to be able to compute the self-loading capacitance and internal power of the gate.

A complete description of substrate capacitances is obtained by analyzing the transistor circuit of each library cell. Figure 9.2 shows a 3-input NOR gate with 3 NMOS transistor in parallel with one end connected to ground and the other end connected to gate output. The PMOS section is made of 3 PMOS transistors in serial connection with one end connected to V_{cc} and the other end connected to the gate output. There are three diffusion capacitances C_0, C_1 and C_2 in this gate. Each of these substrate capacitances have different charging and discharging functions. To charge C_1 for example, it is required that tr_1 be conducting so that a circuit path between V_{cc} and the capacitance is established. To discharge the capacitance, both tr_2 and tr_3 have to be on and at least one of the transistors in the NMOS network has to be on so that a circuit path from the capacitance to ground is established. In summary, to completely specify the diffusion capacitances, we not only need the capacitance value, but also the charging and discharging function associated with each capacitance. The charging and discharging function can be found by analyzing the switch network of the library cell. The self-loading capacitance for a gate is also computed using the diffusion capacitance information.

The *icaps* file format (see appendix B) is used to supplement the capacitance information provided in *genlib* files (SIS library format).

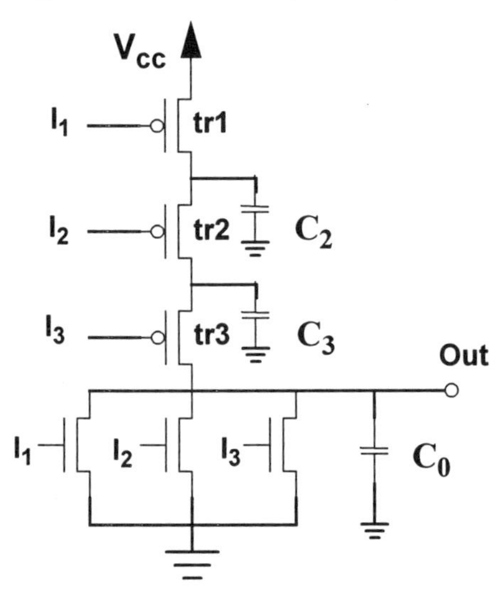

Figure 9.2 Substrate Capacitances for a 3 input NOR Gate

9.3 Power Estimation

9.3.1 Computing power under a zero-delay model

Chapter 2 presented a number of different load models that are used to estimate power during technology independent phase of optimization. A number of issues need to be considered when computing the signal probability of an internal node. Accuracy-speed trade-offs for computing node signal probability values are discussed in the following. The impact of different logic synthesis algorithms on the signal probability and, therefore, the switching activity of nodes in a network is also discussed.

9.3.1.1 Speed/accuracy trade-offs

The immediate fanins of an internal node are generally spatially correlated[2]. This correlation may be present even if primary inputs are spatially uncorrelated. Under these conditions, the spatial correlation at the immediate fanins of a node *n* is due to the reconvergant fanout regions in the transitive fanin cone of *n*. This means that an exact calculation of the signal probability for an internal node *n* requires that this signal probability be computed using the global function of *n*. BDDs have provided a more feasible approach for representing the global function of nodes in a Boolean network. An efficient procedure for computing the signal probability of a function from its BDD representation is also presented in [2]. Therefore BDD based techniques, are good candidate for computing the signal probability of nodes in a network.

Representing the global function of nodes in some circuits may however become too expensive even when BDDs are used to represent the global functions. Therefore it is necessary to provide a mechanism for making speed-accuracy trade-offs when computing signal probabilities. In the following we describe and justify our technique for speeding up the procedure for computing these signal probability values.

---

[2] Spatial correlation between nodes x and y in a network means that the values of x and y in the same clock cycle are independent.

9.3.1.2 Computing signal probabilities using semi-local BDDs

Local BDD for a node n is defined as a BDD where immediate fanins of node n are used in building the BDD. The *Semi-Local* BDD for a node n is defined as the BDD where the nodes used as the BDD variables create a cut in the transitive fanin cone of n. Note that the global and local BDD for a node are special cases of the semi-local BDDs for that node.

The main reason for using global BDDs when computing the signal probability for a node n is to take into account the spatial correlation at the immediate fanins of the node. This correlation is due to reconvergant fanout regions in the transitive fanin cone of node n. Figure 9.3 shows a shallow and a deep reconvergant fanout region in the fanin cone of node n. A shallow reconvergant region spans a less number of levels than a deep reconvergant region. It can be stated that a shallow reconvergant region, generally, results in more spatially correlated fanins to a node. At the same time, a deep reconvergant region will in general result in lower spatial correlation at the immediate fanins of node n. This means that it is possible to capture most of the spatial correlation at immediate inputs of a node n by using the semi-local BDD for that node. The semi-local BDD will take into account the spatial correlation due to shallow reconvergant region while ignoring the spatial correlation due to deep reconvergant fanout regions. This means that using semi-local BDDs allow us to account for most of the spatial correlation at the immediate fanins of a node without building the global BDD for that node. The use of semi-local BDDs for signal probability calculation is discussed in detail in [3].

The definition for deep and shallow regions is implementation dependent. However, experimental results show that in a network that is decomposed into 2-input gates, taking into account Reconvergant regions that span over 5 levels capture a significant part of the spatial correlation at the input of nodes in the network. In POSE,

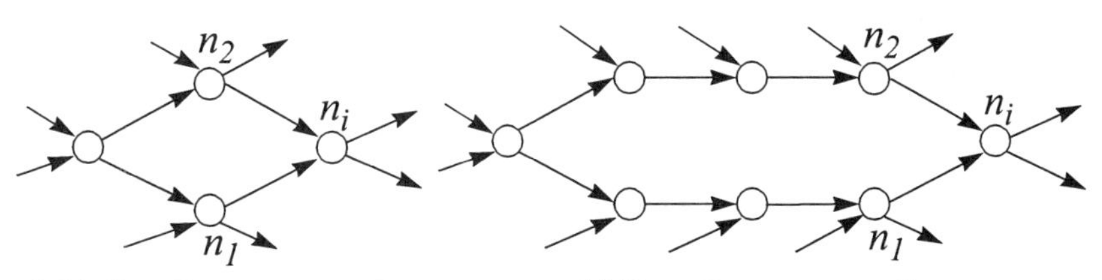

Figure 9.3 Reconvergent Fanout Regions

an option is available for using semi-local BDDs for computing node signal probabilities.

As the network is being optimized, the network structure is modified and, therefore, the definition of semi-local bdds for each node will change. Therefore, if an optimization step changes the network structure drastically, the node switching activity values should be recomputed so that the new activity values reflect the new structure of the network.

It should also be noted that for small networks, using global BDDs provides a more efficient and exact technique for computing the signal probabilities. This is mainly due to the overhead of computing the semi-local BDDs for each node. For larger size networks, however, the use of semi-local BDDs will result in significant speed up during power estimation.

9.3.2 Computing switching activities under a real-delay model

A *tagged probabilistic simulation* approach is described in [7] that correctly accounts for reconvergant fanout and glitches (which are caused by the real delay characteristic of gates). The key idea is to break the set of possible logical waveforms at a node *n* into four groups, each group being characterized by its steady state values (i.e., values at time instance 0^- and •). Next, each group is combined into a probability waveform with the appropriate steady-state tag (see Figure 9.4). Given the tagged probability waveforms at the input of a simple gate, it is then possible to compute the tagged probability waveforms at the output of the gate. The correlation between probability waveforms at the inputs is approximated by the correlation between the steady state values of these lines. This is much more efficient than trying to estimate the

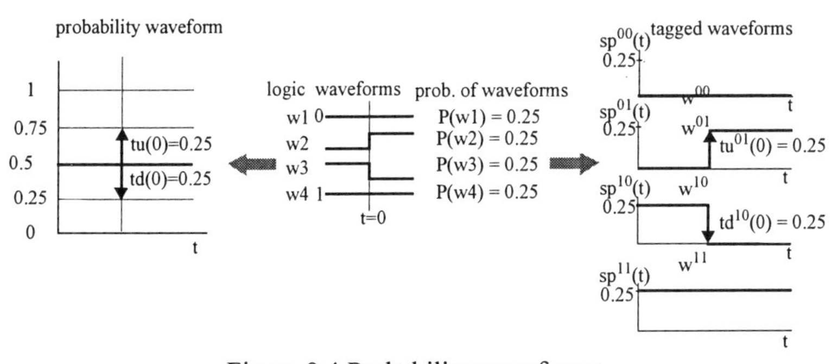

Figure 9.4 Probability waveforms

dynamic correlations between each pair of events. This approach requires significantly less memory and runs much faster than symbolic simulation, yet achieves very high accuracy, e.g., the average error in aggregate power consumption is about 10%. In order to achieve this level of accuracy, detailed timing simulation along with careful *glitch filtering* and *library characterization* are needed [3]. The first item refers to the scheme for eliminating some of the short glitches that cannot overcome the gate inertias from the probability waveforms. The second item refers to the process of generating accurate and detailed macro-modeling data for the gates in the cell library.

9.3.3 Effect of optimization on switching activities

A correct methodology for low power synthesis requires that the signal probability and switching activity values computed and stored on each node remain correct as the network is being restructured during the logic synthesis process. Most operations on a Boolean network are algebraic in nature. This means that global function of nodes and therefore the zero-delay signal probability of nodes do not change for these operations. Some logic synthesis operations, however, do modify the global function of the node being optimized and other nodes in the network. These operations will result in changing the signal probability of nodes in the network without having modified these nodes. Figure 9.5 shows an example where simplifying the function of a node using its observability don't care has resulted in changing the signal probability of the node and nodes in its transitive fanout.

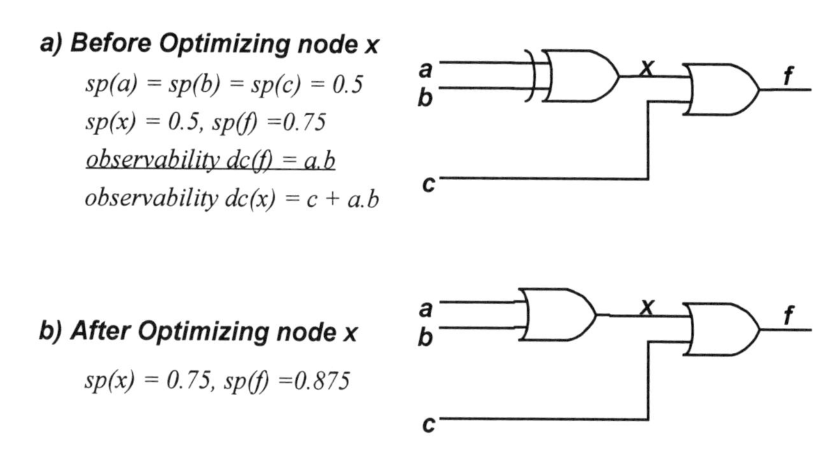

Figure 9.5 Effect of optimizations on signal probabilities

If switching activity values are estimated using a real delay model, then after changing a node *n*, the delay of the network changes and, therefore, the switching activity of all nodes in transitive fanout of *n*, and the switching activity of all nodes in the transitive fanouts of the immediate fanins of node *n* change. This is because changing the function of a node changes the load on its immediate fanins, which changes the delay for fanout nodes of the immediate fanin nodes.

9.4 Power Estimation Flow

The following example illustrates a typical session for performing power estimation in the POSE environment. The following assumes that the appropriate directories for *blif* and *sw* files have been included in the *open_path* environment variable of POSE.

```
alburz.usc.edu(408)pose
USC POSE Development Version (compiled 26-Sep-95 at 4:50 PM)
pose> read_blif C17.blif
pose> sw_read pose/ex/sw/sw01/C17.sw
pose> dls_trace -s -L sop
pose> sw_print -p
-----------------------------------------------------------------
Activity trace has been started.
      dl_model: SOP form.
-----------------------------------------------------------------
         Node Name  SigProb   SwAct  Out_Load  Int_Load
-----------------------------------------------------------------
-PI-       1GAT(0)   0.883   0.207   1.000   0.000
-PI-       2GAT(1)   0.690   0.428   1.000   0.000
-PI-       3GAT(2)   0.244   0.369   2.000   0.000
-PI-       6GAT(3)   0.595   0.482   1.000   0.000
-PI-       7GAT(4)   0.410   0.484   1.000   0.000
        {22GAT(10)}  0.745   0.380   0.000   0.000
        {23GAT(9)}   0.698   0.421   0.000   0.000
          11GAT(5)   0.855   0.248   2.000   0.000
          10GAT(6)   0.784   0.338   1.000   0.000
          19GAT(7)   0.649   0.455   1.000   0.000
          16GAT(8)   0.411   0.484   2.000   0.000
-----------------------------------------------------------------
pose> so pose/pose_lib/sc.stats
```

```
XXXXXXXXXXXXXXXXXXXXXXXXXXXXXXXXXXXXXXXXXXXXXXXXXXXXXX
C17.iscas     pi= 5  po= 2  nodes= 6      latches= 0
lits(sop)= 12  lits(fac)= 12
Net stats ignoring primary input and outputs:
avg node size = 1.500000
stddev node size = 0.866025
min node size = 0
max node size = 2
avg node switching activity = 0.391128
stddev node switching activity = 0.068957
min node sa = 0.248255
max node sa = 0.484017
=-=-=-=-=-=-=-=-=-=-=-=-=-=-=-=-=-=-=-=-=-=-=-=-=-=-=-=-=-=-=-=-=-=-=
Previous sp trace on this network is still active.
L:sop:Total Area(SOP) = 12
L:sop:Total Area(FAC) = 12
L:sop:Total Load     = 12.000000
L:sop:Total Activity  = 4.297156
L:sop:Total Power     = 4.596846
=-=-=-=-=-=-=-=-=-=-=-=-=-=-=-=-=-=-=-=-=-=-=-=-=-=-=-=-=-=-=-=-=-=-=
Previous sp trace on this network is still active.
L:a_factor:Total Area(SOP) = 12
L:a_factor:Total Area(FAC) = 12
L:a_factor:Total Load     = 12.000000
L:a_factor:Total Activity  = 4.297156
L:a_factor:Total Power     = 4.596846

XXXXXXXXXXXXXXXXXXXXXXXXXXXXXXXXXXXXXXXXXXXXXXXXXXXX
pose> read_library -n pose/pose_lib/bam.genlib
pose> sw_read_icaps pose/pose_lib/bam.capinfo
pose> map -x -m pd0 -s
WARNING: uses as primary input drive the value (3.50,2.20)
WARNING: uses as primary input arrival the value (0.00,0.00)
WARNING: uses as primary input max load limit the value (999.00)
WARNING: uses as primary output required the value (0.00,0.00)
WARNING: uses as primary output load the value 0.04
total switching rate: 4.98 power (weighted switching): 0.19
total switching rate: 4.98 energy (power-delay product): 0.19
total gate area: 12988.00 maximum arrival time: 2.9962
CPU time for mapping: 0.183 s
pose> sw_est
Total Load     = 0.504000
Total Activity  = 4.981695
Total Power     = 0.190315
```

```
pose> sw_print -p
-----------------------------------------------------------------
Activity trace has been started.
Network is mapped.
-----------------------------------------------------------------
        Node Name   SigProb    SwAct   Lib_Load
-----------------------------------------------------------------
-PI-       1GAT(0)    0.883    0.207    0.040
-PI-       2GAT(1)    0.690    0.428    0.021
-PI-       3GAT(2)    0.244    0.369    0.071
-PI-       6GAT(3)    0.595    0.482    0.033
-PI-       7GAT(4)    0.410    0.484    0.040
             [559]    0.310    0.428    0.029
          11GAT(5)    0.855    0.248    0.061
             [572]    0.145    0.248    0.031
             [579]    0.589    0.484    0.060
             [568]    0.255    0.380    0.021
       {22GAT(10)}    0.745    0.380    0.038
             [570]    0.302    0.421    0.021
        {23GAT(9)}    0.698    0.421    0.038
-----------------------------------------------------------------
pose> so pose/pose_lib/sc.map.stats
XXXXXXXXXXXXXXXXXXXXXXXXXXXXXXXXXXXXXXXX
Previous sp trace on this network is still active.
Mapped:Total Load      = 0.504000
Mapped:Total Activity  = 4.981695
Mapped:Total Power     = 0.190315
=-=-=-=-=-=-=-=-=-=-=-=-=-=-=-=-=-=-=-=-=-=-=-=-=-=-=
aoi21       :  2 (area=2448.00)
inv1x       :  4 (area=1156.00)
nand2       :  1 (area=1768.00)
nor2        :  1 (area=1700.00)
Total: 8 gates, 12988.00 area
=-=-=-=-=-=-=-=-=-=-=-=-=-=-=-=-=-=-=-=-=-=-=-=-=-=-=
Net stats ignoring primary input and outputs:
avg node size = 1.400000
stddev node size = 1.019804
min node size = 0
max node size = 3
avg node switching activity = 0.381356
stddev node switching activity = 0.072786
min node sa = 0.248255
max node sa = 0.484017
C17.iscas     pi= 5   po= 2   nodes=  8      latches= 0
```

```
lits(sop)= 16 lits(fac)= 14
XXXXXXXXXXXXXXXXXXXXXXXXXXXXXXXXXXXXXXXXXXXXXXX
pose> quit
alburz.usc.edu(409):
```

9.5 Experimental Results

POSE has been implemented and results have been generated for minimum power circuits. Table 9.1 gives the power consumption of multi-level circuits in the MCNC benchmark set when they are optimized and mapped for minimum area, and when they are optimized and mapped for minimum power. Columns 2, 3, and 4 give the area, delay, and power (switched capacitance) of the technology mapped circuits when they are optimized for minimum area. Columns 5, 6, and 7 give the area, delay and power (normalized with respect to values in the first three columns) for the same circuits when they are optimized and technology mapped for minimum power consumption using the methodology presented in section 9.1. As the results show, POSE has been able to reduce the circuit power consumption on average by 26% at the expense of increasing the area of the circuits by 23%. This clearly shows a trade-off between area and power. The delay of the circuits has only been increased by 4%. Table 9.2 gives the same statistics for two-level example benchmarks in the MCNC benchmark set. The power consumption of the technology mapped circuits has been reduced on average by 29% at the expense of increasing the area on average by 30%. Again, this clearly shows that POSE has been able to successfully find the minimum power solution by trading off area for power.

Ex	Minimum Area			Minimum Power		
	area	delay	power	area	delay	power
C1355	389760	24.38	16.08	1.12	1.16	0.93
C1908	434768	39.02	14.78	1.12	1.13	0.84
C432	170288	43.54	6.65	1.20	0.97	0.84
C5315	1365088	41.50	65.89	1.25	0.99	0.85
alu2	305312	40.43	9.67	1.19	1.20	0.64
alu4	592528	48.71	9.83	1.43	1.17	0.70
b9	110896	9.34	4.17	1.13	1.09	0.78
dalu	760960	50.26	23.75	1.26	1.06	0.68
des	2865200	162.98	111.72	1.34	0.78	0.81
frg1	115536	18.92	4.98	1.25	0.86	0.71
frg2	692752	36.38	19.76	1.18	1.13	0.68
i8	797616	39.83	20.46	1.35	0.95	0.68
k2	1011056	32.84	12.29	1.44	1.13	0.65
rot	594848	26.13	22.84	1.20	1.35	0.78
sct	75168	31.43	2.51	1.00	0.44	0.60
t481	612944	30.85	7.89	1.15	0.90	0.68
term1	149408	13.49	5.17	1.03	1.10	0.71
ttt2	193952	17.71	6.32	1.01	0.99	0.61
9symml	159152	22.42	6.99	1.68	1.26	0.77
Average				1.23	1.04	0.74

Table 9.1 POSE: Minimum power solutions for multi-level benchmarks

Ex	Minimum Area			Minimum Power		
	area	delay	power	area	delay	power
apex1	1308944	30.42	31.67	1.33	1.34	0.68
apex5	676512	34.21	12.92	1.40	0.70	0.72
b12	75168	10.96	2.19	1.35	1.10	0.85
bw	130848	33.13	4.41	1.49	0.49	0.67
clip	118320	20.30	5.60	1.17	1.17	0.71
cps	1050960	36.72	22.69	1.34	1.07	0.58
duke2	392544	33.08	10.19	1.13	0.72	0.61
e64	293248	109.49	4.22	1.00	1.01	0.63
ex4	402288	12.37	15.97	1.39	1.26	0.83
misex1	46864	13.54	1.70	1.43	0.80	0.81
misex2	91872	10.87	2.52	1.34	1.07	0.85
misex3	575360	32.13	15.65	1.40	1.20	0.69
pdc	341040	21.25	12.08	1.40	1.06	0.83
rd73	58928	17.85	1.35	1.25	0.98	0.68
rd84	128992	18.21	4.70	1.24	1.33	0.65
spla	545200	24.23	15.81	1.35	1.15	0.62
vg2	85376	11.00	3.42	1.24	1.47	0.71
5xp1	102544	30.88	4.05	0.99	0.53	0.66
9sym	178640	19.08	9.06	1.48	0.96	0.78
Average				1.30	1.02	0.71

Table 9.2 POSE: Minimum power solutions for two-level benchmarks

Appendix A: Input Activity File Format

The following format called the swfile format, is used for input/output of switching information in POSE.

For each primary input of the network, the sw_file will contain a line in the following format:

.sp input_name signal_probability

The following is file C17.sw which stores the signal probability information for file C17.blif.

.sp 1GAT(0) 0.065046
.sp 2GAT(1) 0.160337
.sp 3GAT(2) 0.853059
.sp 6GAT(3) 0.145651
.sp 7GAT(4) 0.012658

The signal probability for all unspecified inputs will be set to a given default value (see sw_read command) and a warning message will be printed.

Appendix B: ICAPS File Format

Two types of capacitances are specified in the *icaps* file: 1-input capacitances and 2-substrate capacitances. Although the input capacitance specification is duplicated in *genlib* file, the rationale here is that the values specified in *genlib* files could have been optimized for delay calculation while in *capinfo* file they are optimized for power estimation. The input capacitances included in the ICAPS file is not used in this release of POSE.

The capacitances in a library cell are specified in the following format:

```
GATE <cell-name>
        input_cap <pin_number> <capacitance_value>
        .
        .
        int_cap <capacitance_value> c <charging_function>
                        d <discharging_function> e
        .
        .
        .
```

The capacitance values are in pF. The *input_cap* specifies the input capacitances. The pin number starts with 0 with the number order matching the pin order in the *genlib* file. The *int_cap* specifies substrate capacitances. The variables used in *charging_function* and *discharging_function* are "*pin_number*"*s*. Moreover the functions are in conjunctive form with a space between two clauses. For example (0+1)(1+2) is specified as 0+1 1+2. A special variable "o", which represents the output function, can be very handy. In the following we show an example for a 2-input NOR gate.

```
GATE NOR2
        input_cap 0 0.111
        input_cap 1 0.111
        int_cap 0.121915 c !0 d !o !1 e
        int_cap 0.076 c o d !o e
```

Both the input capacitances for the first pin and second pin (which refers to the corresponding *genlib* file specification) are 0.111 pF. The first substrate capacitance is the one between the two PMOS transistor in this gate. Its value is 0.121915pF. To charge this capacitance, the upper PMOS transistor has to be conducting. In this case, the gate input is pin 0. Therefore the charging function is !0. To discharge the capacitance, the lower PMOS transistor has to be ON and at least one of the NMOS transistors also has to be ON. That is, the discharging function is !(0+1)*(!1)=(!o)(!1). The last substrate capacitance is the output capacitance and its charging and discharging functions are o and !o, respectively.

Appendix C: Extended Commands

Extended commands are used in POSE to augment the existing commands with power relevant options. The idea behind extended commands is that by specifying an option on the command line, a different procedure will be executed for that same command. This allows the extension of current commands with power options. For example, running "gkx" will perform regular algebraic extraction procedure. However, running "gkx -p" will perform algebraic extraction for low power without the original code for "gkx" having been modified.

A list of extended commands and the command line options that activate the power relevant procedures is obtained by using the command "help -e".

```
alburz(442):pose
USC, POSE Development Version (compiled 7-May-95 at 10:12 PM)
pose> help -e
(decomp      -p)    (eliminate     -p)    (factor      -p)
(gcx         -p)    (gkx           -p)    (help        -e)
(map         -x)    (print_value   -e)    (simplify    -p)
pose>
```

To see the command line options for extended "gkx" and "print_value" command type:

```
pose> gkx -p -h
pose> print_value -e -h
```

Note: Experiments show that for a network with a higher number of kernels, low power extraction techniques usually produce better results. The extended simplify command is used to prevent the simplification step from substituting smaller nodes into other nodes. This means that the network area will not be minimized maximally (therefore leaving more freedom for low power extraction step). However, this loss in area will be small since only small nodes have been prevented from being substituted. The definition of "small" nodes is given as an option to the extended "simplify" command.

Appendix D: POSE Commands

The following commands are used to perform the basic steps in performing power analysis. In addition to these new commands, some of the existing commands in SIS have been extended to consider power consumption. A list of extended commands is obtained by typing *"help -e"*.

sw_set_in [options]
This command is used to set the signal probabilities of the network primary inputs. With no arguments, the signal probability of all primary inputs is set to 0.5.

Arguments:

-i	print the input signal probabilities
-s <value>	set the input signal probabilities to the specified <value>
-r	set the input signal probabilities to random values between [0.0,1.0]
-e	unset input signal probabilities
-o <value>	set the load on circuit outputs as an integer multiple of internal line caps. This information is only relevant during technology independent optimization.

sw_read [options] filename
This command is used to read in the signal probability information from the file "filename" which is assumed to be in *swfile* format. Usual filename conventions apply: - (or no filename) stands for standard input, and tilde- expansion is performed on the filename.Normal operation is to replace the current probability information with a new values. If the signal probability is not specified for any inputs, then the signal probability for that input is set to a default value or a random value based on the command line options. A warning message is also printed in this case.

Arguments:

-s <value>	set the input signal probability of all unspecified inputs to <value>. If <value>=-1 then set the signal probabilities to random values between [0.0,1.0]

sw_write filename

Write the current signal probabilities to file filename in the swfile format. Usual filename conventions apply: - (or no filename) stands for standard output, and tilde-expansion is performed on the "filename".

Arguments:

none

dls_trace [options]

This command is used to set the load, delay and switching activity model options and also perform switching activity trace on the network. *dls_trace* should be used before commands that require load or switching activity information. This command is only used to set up the structures or options that are required to perform load, delay and activity trace and maintain this information. Therefore this is a fast command. The actual values for load and switching activity will be computed on-demand as they are requested by the optimization/estimation commands which follow the trace command.

With no arguments, this command does not perform any operations.

Arguments:

-m <value>	limit the number of nodes in the bdd package to 2^value
-L <option>	Use the given load model in estimating delay/loads:
a_factor	use factored form for area model
p_factor	use factored form for power model
simple	use simple node model
sop	use SOP form model
-M opt	set default switching trace method:
global	use global bdds (def)
local	use semi-local bdds
-N <value>	number of levels to use for local bdds(def.N=5)
-S <value>	sw trace strategy(sw_trace is NOT restarted):
1	ALWAYS_VALID: Never invalidate node bdd or switching activity(def)
2	TFO_BDD_SW_INVALID: Invalidate the bdd and switching activity of node and all its transitive fanouts
-s	start switching activity trace

-e end switching activity trace

Note that changing the Load trace option will automatically invalidate all previous load/delay trace performed using another option.

sw_read_icaps filename

This command is used to read an information file for the internal capacitances of the gates in the library. This information is used to more accurately compute the output load of gates in a mapped network. After reading the library using the "read_library" command, read the appropriate icaps file (provided in pose_lib directory) and all power estimations on the mapped network will include the icaps informations. If no icaps file is read, then load is computed ignoring the icaps information.

sw_est [options]

This command uses the current switching activity and load/delay options to report power of the network. It reports the total load, total switching activity and total power of the network. Internal load is assumed to be zero when the network is mapped. The load model used for power estimation is adopted from the load/delay option given using *dls_trace* command. Power is always reported as the sum of switched-capacitance for all nodes in the network. The scaling factors due to supply voltage and clock frequency are not considered.

Arguments:

-f	Print fast statistics on power distribution of the network.
-w	Consider wiring load in load calculations when the network is mapped. This wiring is estimated using the delay package
-c	Consider self-loading caps specified in icaps file. (if icaps file is not read, then this flag does not do anything)
-i	Do not report internal power for unmapped networks

sw_print [options]

This function is used to report different network statistics; This command can also be used to print a histogram of the distribution of the switching activity for different load values.

Arguments:

-n	print average node size and switching activities
-p	print a history of current traces, trace options, and activity and load values for the nodes
-f	report power in the factored form of the network. This value is actually the sum of literals in the factored form of nodes weighted by their switching activities.
-h	print a histogram of load versus switching activity
-I <value>	specify the number of increments on the histogram
-M <value>	Specify the maximum value on the vertical axis. (default from data)
-N	do not include primary input nodes in the histogram

sw_decomp

This function is used to decompose each node in the network in the SOP form. Each cube of the node is changed into a node and the node is changed to an OR gate. This procedure does not decompose XOR or XNOR nodes. This step is used after the network is optimized for power and allows for further reduction in power by performing power optimization on the decomposed network.

References

[1] "SIS: A system for sequential circuit synthesis," Report M92/41, UC Berkeley, 1992.

[2] S. Chakravarty. "On the complexity of using BDDs for the synthesis and analysis of boolean circuits." In proceedings of the *27th Annual Allerton Conference on Communication, Control and Computing*, pages 730–739, 1989.

[3] C-S. Ding and M. Pedram. "Tagged probabilistic simulation provides accurate and efficient power estimates at the gate level." In proceedings of the *Symposium on Low Power Electronics*, September 1995.

[4] S. Iman and M. Pedram. "Logic extraction and factorization for low power." In proceedings of the *ACM/IEEE Design Automation Conference*, June 1995.

[5] R. Marculescu, D. Marculescu, M. Pedram. "Logic level power estimation considering spatio-temporal correlations." In proceedings of the *IEEE International Conference on Computer Aided Design*, pages 294-299, 1994.

[6] R. Rudell. "Logic Synthesis for VLSI Design." Ph.D. thesis, University of California, Berkeley, 1989.

[7] C-Y. Tsui, M. Pedram, C-H. Chen, and A. M. Despain. "Low power state assignment targeting two- and multi-level logic implementations." In proceedings of the *IEEE International Conference on Computer Aided Design*, pages 82–87, November 1994.

[8] C-Y. Tsui, M. Pedram and A. M. Despain. "Power efficient technology decomposition and mapping under an extended power consumption model." In *IEEE Transactions on Computer Aided Design*, Vol.~13, No.~9 (1994), pages 1110--1122.

Part V

Conclusion

CHAPTER 10

Concluding Remarks

The main contribution of this book has been to develop the techniques and the methodology for effective power reduction during logic synthesis. A necessary requirement for effective power optimization is a power model that can effectively be used during logic synthesis. This book also presented a power model which was proven to be highly accurate for minimizing the zero-delay model power cost. POSE (Power Optimization and Synthesis Environment) has been developed by implementing the techniques described in this book. POSE provides the necessary steps for transforming the Boolean representation of a design into a technology mapped circuit. POSE also provides the necessary power estimation techniques to guide the power optimization procedures.

Experimental results presented in chapter 2 demonstrate the accuracy of the power models which are used to guide the technology independent optimization steps. It was demonstrated that the combination of more accurate load models, and the choice of zero-delay switching activity estimates provide a tangible power model that closely reflects the power consumption of the mapped circuit. Techniques and methodologies implemented within the POSE system use this power model to minimize the power consumption of the Boolean network. Results obtained using POSE show that, on average, the power consumption of example benchmarks are reduced by %26 at the expense of increasing area by %23 when starting with multilevel circuits. Power consumption is reduced by %29 at the expense of increasing area by %30 when starting with two-level circuits. It is also shown that, on average, the delay of these circuits is not changed.

10.1 Conclusions

The following sections provide concluding remarks on the results of each chapter in this book.

10.1.1 Two-Level Logic Minimization in CMOS Circuits (chapter 3)

This chapter presented an exact algorithm for minimizing the power consumption of a function in the sum of the products form. An analysis based on the statistical properties of the inputs was also presented to give insight into the increased complexity of solving the minimum power problem. This analysis was then used to propose methods for controlling the increased complexity of the minimum covering problem. The method presented in this chapter has been implemented, and results are reported for a set of experiments on random functions of 3 to 9 variables.

10.1.2 Two-Level Logic Minimization in PLAs (chapter 4)

This chapter presented a solution to the problem of minimizing power consumption in PLAs. In doing so, it was shown that for dynamic PLA circuits, as with a minimum area solution, a minimum power solution also consists only of prime implicants of the function. Therefore, approaches used for minimizing area of PLAs may also be used to minimize the PLA power consumption. The same results were also proved for pseudo NMOS implementations of incompletely specified single output functions and completely specified multiple output functions. A heuristic approach was also presented for pseudo-NMOS implementations of incompletely specified multiple output functions. The results in this chapter also show that during function minimization for dynamic CMOS PLA, it is still possible to make area-power trade-offs by using implicants with fewer literals, and also literals with input statistics that result in a lower power consumption. This reduction in power, however, has been at the expense of increasing the number of implicants in the final cover of the function which will in turn result in an increase in the number of rows of the PLA and, therefore, an increase in area.

10.1.3 Logic Restructuring for Low Power (chapter 5)

This chapter introduced a unified approach to power optimization using algebraic based methods. The results show that it is generally possible to slightly increase the network area in order to reduce the power consumption of the technology mapped network. This is accomplished by reducing load on high activity nets and by introducing new nodes which have a lower switching activity. Power cost functions were proposed to find the quality of extractions performed for low power, which proved to be quite effective.

10.1.4 Logic Minimization for Low Power (chapter 6)

In this chapter a method is presented that allows for minimizing the power consumption of a network using local don't care conditions and local function minimization. Using the techniques presented here, it is possible to guarantee that local node optimization will degrade the global power consumption of the network. This means that local nodes can be optimized without concern for how changes in the function of the current node affect the power consumption in the rest of the network. A method for optimizing the local function of node was presented where the concept of minimal literal and variable supports have been used to find the lowest cost input supports and then lowest cost implementation of the function.

10.1.5 Technology Dependent Optimization for Low Power (chapter 7)

In this chapter techniques for performing technology decomposition and mapping for low power are presented. The goal for technology decomposition technique is to minimize the total switching activity of the decomposed logic. The goal during technology mapping is to reduce power by hiding high activity nodes inside complex library gates. The technology mapping algorithm uses a dynamic programming approach where power-delay trade-off curves are generated for each node in the technology decomposed network in a depth-first order. The best gate matching at each node which meets the timing constraints is then selected from the non-inferior points on power-delay curve for each node. This chapter also presents a gate-resizing technique which takes advantage of the slack available at the output of gates in the technology mapped network. By re-sizing the gates which are not on the critical path, power is further reduced.

10.1.6 Post Mapping Structural Optimization for Low Power (chapter 8)

In this chapter a technique for performing post-mapping optimization for low power is presented. This technique is based on structural transformation which leads to reducing power in the mapped network. This technique is based on replacing a single wire in the circuit with another, or a combination of other existing wires in the circuit. A power model is presented which accurately predicts the change in power for each possible wire replacement. This power model is then used to guide the optimization process. Experimental results show that, on average, power is reduced by %10 using this optimization technique.

10.1.7 POSE: Power Optimization and Synthesis Environment (chapter 9)

This chapter presented a methodology for designing low power digital circuits. POSE has been developed as a unified framework for specifying and maintaining power relevant circuit information. POSE also provides power estimation and optimization procedures which facilitate the low power design methodology.

The emphasis in this chapter has been in identifying an effective low power design methodology. In addition, POSE has been developed and tested with the proposed methodology where it is shown that the power estimation techniques provided a good method for verifying the quality of the incrementally power optimized circuits. An optimization script has also been developed based on the methodology presented in this chapter.

POSE is a first step in developing a complete environment for low power circuit synthesis. Many optimization algorithms at different levels of abstraction are yet to be incorporated into the POSE environment. At the same time, power estimation techniques should also be included at different levels of the design abstraction. In accordance with this philosophy, POSE has been developed to allow researchers to easily incorporate new power estimation and optimization techniques into POSE. This will then allow POSE to be used as a platform for future research on low power CAD tools. The first release of POSE has been made[1] available and is being used as both an optimization/estimation tool and as a platform for implementing and experimenting with new low power design techniques.

10.2 Future Directions

The main goal in the design of digital circuits is to maximize circuit speed while minimizing cost. This goal has traditionally translated into a trade-off between circuit

[1]. POSE can be obtained on the World Wide Web at http://atrak.usc.edu/~pose.

delay and area. With the increasing popularity of low power circuits, power has become the third major cost to be minimized during the design process. With the introduction of power as a major cost function, it is now necessary to study the area-power and delay-power trade-offs.

The work presented in this book has concentrated on minimizing the power consumption of the design, at the same time attempting to provide a low area solution. The incentive for this approach is the observation that power tracks well with area and, therefore, the area for a low power design should not be increased drastically.

Future research in logic synthesis needs to study the interactions between different costs being optimized. The following provides a discussion on these trade-offs.

10.2.1 Area-Power Trade-Offs

It is generally accepted that during logic synthesis, power tracks well with area. This means that a larger design will generally consume more power. It is, however, possible to make area-power trade-offs during logic synthesis. The effect of these trade-offs can be seen in the final implementation (tables 9.1 and 9.2) where the circuit power is reduced, but its area may be more than the area of the same design optimized for minimum area. Note that in both power optimized and area optimized circuits, area is reduced during the synthesis process. However, in a power optimized design, area is not reduced as much as the area optimized design, therefore, leading to an area-power trade-off. As can be seen in the results generated using the POSE system, the power is, on average, reduced by %30, where area is increased on average by %30. Power density (defined as ratio of power to area) is a measure that can be used to study the effectiveness of a power optimization process. Note for the designs generated using POSE, the power density has actually been reduced by %50. However, this reduction in power density is offset by the increase in the circuit area. Additionally, the increase in circuit area will result in more standby currents which will in turn increase the power. Therefore, it is desired to develop power optimization methodologies where power is reduced with only a small increase the circuit area. Improvements in this direction will allow even more power reduction during the logic synthesis phase of the design.

10.2.2 Delay-Power Trade-Offs

The design constraints are usually specified in terms of achieving a given throughput, while minimizing the cost. For low power designs, cost is defined as the power consumption of the circuit, therefore, delay constrained power optimization is a necessary part of the design cycle. The techniques developed in this thesis can all be applied during delay constrained power optimization process. The technology mapping algorithm implemented within POSE is also capable of performing delay constrained power minimization. The main challenge here is to develop a methodology to effectively perform delay constrained power optimization using the existing optimization techniques. For example, the main flow for delay constrained area optimization is to first optimize the circuit for minimum area and then recover delay by restructuring the design using collapsing and partitioning techniques. For delay constrained power optimization, a combination of area optimization, power optimization, and circuit re-structuring techniques need to be developed which leads to maximum power reduction subject to a given delay.

Another issue in delay-power trade offs is the goal to minimize energy versus the goal to minimize power. For example, for portable devices, the goal is to minimize energy since the energy source (i.e. the battery) has a limited supply. At the same time, the goal in VLSI chip designs is to minimize power since the generated heat is limiting the device density. Future research should concentrate on developing techniques to address energy versus power optimization. This problem is particularly interesting since the power consumption is a function of the circuit delay since the clock frequency is decided by the maximum delay in the circuit. At the same time, the energy consumption is a function of the number of operations performed in the circuit as opposed to how fast these operations are performed. The inherent difference in these two optimization problems brings forth a new set of challenges that need to be addressed in future research directions.

Index